MEASURING ACCESS

TO LEARNING OPPORTUNITIES

Committee on Improving Measures of Access to
Equal Educational Opportunity

Willis D. Hawley and Timothy Ready, *Editors*

Center for Education and Committee on National Statistics

Division of Behavioral and Social Sciences and Education

NATIONAL RESEARCH COUNCIL
OF THE NATIONAL ACADEMIES

THE NATIONAL ACADEMIES PRESS
Washington, D.C.
www.nap.edu

THE NATIONAL ACADEMIES PRESS 500 Fifth Street, N.W. Washington, DC 20001

NOTICE: The project that is the subject of this report was approved by the Governing Board of the National Research Council, whose members are drawn from the councils of the National Academy of Sciences, the National Academy of Engineering, and the Institute of Medicine. The members of the committee responsible for the report were chosen for their special competences and with regard for appropriate balance.

This study was supported by Contract/Grant No. R215U990016-01B between the National Academy of Sciences and the U.S. Department of Education. Any opinions, findings, conclusions, or recommendations expressed in this publication are those of the author(s) and do not necessarily reflect the views of the organizations or agencies that provided support for the project.

Library of Congress Cataloging-in-Publication Data

National Research Council (U.S.). Committee on Improving Measures of
Access to Equal Educational Opportunity.
 Measuring access to learning opportunities / Committee on Improving
Measures of Access to Equal Educational Opportunity ; Willis D. Hawley
and Timothy Ready, editors.
    p. cm.
"Center for Education and Committee on National Statistics, Division of
Behavioral and Social Sciences and Education."
Includes bibliographical references.
 ISBN 0-309-08897-6 (pbk.)
 1. Educational equalization—United States. 2. Minority
students—Civil rights—United States. 3. Educational surveys—United
States. I. Hawley, Willis D. II. Ready, Timothy. III. Title.
 LC213.2.N396 2003
 379.2′6′0973—dc21
                                    2003007413

International Standard Book Number 0-309-50539-9 (PDF)

Additional copies of this report are available from the National Academies Press, 500 Fifth Street, N.W., Lockbox 285, Washington, DC 20055; (800) 624-6242 or (202) 334-3313 (in the Washington metropolitan area); Internet, http://www.nap.edu

Suggested citation: National Research Council. (2003). *Measuring access to learning opportunities.* Committee on Improving Measures of Access to Equal Educational Opportunity. W.D. Hawley and T. Ready, editors. Center for Education and Committee on National Statistics, Division of Behavioral and Social Sciences and Education. Washington, DC: The National Academies Press.

THE NATIONAL ACADEMIES
Advisers to the Nation on Science, Engineering, and Medicine

The **National Academy of Sciences** is a private, nonprofit, self-perpetuating society of distinguished scholars engaged in scientific and engineering research, dedicated to the furtherance of science and technology and to their use for the general welfare. Upon the authority of the charter granted to it by the Congress in 1863, the Academy has a mandate that requires it to advise the federal government on scientific and technical matters. Dr. Bruce M. Alberts is president of the National Academy of Sciences.

The **National Academy of Engineering** was established in 1964, under the charter of the National Academy of Sciences, as a parallel organization of outstanding engineers. It is autonomous in its administration and in the selection of its members, sharing with the National Academy of Sciences the responsibility for advising the federal government. The National Academy of Engineering also sponsors engineering programs aimed at meeting national needs, encourages education and research, and recognizes the superior achievements of engineers. Dr. Wm. A. Wulf is president of the National Academy of Engineering.

The **Institute of Medicine** was established in 1970 by the National Academy of Sciences to secure the services of eminent members of appropriate professions in the examination of policy matters pertaining to the health of the public. The Institute acts under the responsibility given to the National Academy of Sciences by its congressional charter to be an adviser to the federal government and, upon its own initiative, to identify issues of medical care, research, and education. Dr. Harvey V. Fineberg is president of the Institute of Medicine.

The **National Research Council** was organized by the National Academy of Sciences in 1916 to associate the broad community of science and technology with the Academy's purposes of furthering knowledge and advising the federal government. Functioning in accordance with general policies determined by the Academy, the Council has become the principal operating agency of both the National Academy of Sciences and the National Academy of Engineering in providing services to the government, the public, and the scientific and engineering communities. The Council is administered jointly by both Academies and the Institute of Medicine. Dr. Bruce M. Alberts and Dr. Wm. A. Wulf are chair and vice chair, respectively, of the National Research Council.

www.national-academies.org

COMMITTEE ON IMPROVING MEASURES OF ACCESS TO EQUAL EDUCATIONAL OPPORTUNITY

WILLIS D. HAWLEY (*Chair*), Department of Educational Policy, University of Maryland

JULIAN BETTS, Department of Economics, University of California, San Diego

JOMILLS H. BRADDOCK II, Department of Sociology, University of Miami

GARNET (LAVAN) DUKES, Florida Department of Education, Tallahassee

JOAN FIRST, National Coalition of Advocates for Students, Boston, MA

JOHN FORREST KAIN, Cecil and Ida Green Center for the Study of Science and Society, University of Texas, Dallas

VALERIE E. LEE, School of Education, University of Michigan

WEI-WEI LOU, Portland Public Schools, Oregon

JENS LUDWIG, Public Policy Institute, Georgetown University

GARY ORFIELD, Harvard Graduate School of Education

TIMOTHY READY, *Study Director*

PASQUALE DE VITO, *Senior Program Officer*

ANDREW TOMPKINS, *Research Assistant*

TERRY HOLMER, *Senior Project Assistant*

Acknowledgments

This report has been reviewed in draft form by individuals chosen for their diverse perspectives and technical expertise, in accordance with procedures approved by the National Research Council's Report Review Committee. The purpose of this independent review is to provide candid and critical comments that will assist the institution in making its published report as sound as possible and to ensure that the report meets institutional standards for objectivity, evidence, and responsiveness to the study charge. The review comments and draft manuscript remain confidential to protect the integrity of the deliberative process. We wish to thank the following individuals for their review of this report: Robert Bell, AT&T Research Laboratories, Florham Park, NJ; Sue Berryman, World Bank, Washington, DC; Tamela Lea Eitle, University of Miami, FL; Scott Palmer, Nixon Peabody LLP, Washington, DC; Dennis Parker, NAACP Legal Defense & Educational Fund, Inc., New York; Paul Smith, Children's Defense Fund, Washington, DC; Karl Taeuber, University of Wisconsin-Madison; William L. Taylor, Law Offices of William L. Taylor, Washington, DC; J. Douglas Willms, University of Brunswick, Fredericton, NB.

Although the reviewers listed above have provided many constructive comments and suggestions, they were not asked to endorse the conclusions or recommendations nor did they see the final draft of the report before its release. The review of this report was overseen by William T. Trent, Department of Educational Policy Studies, University of Illinois-Champaign. Appointed by the National Research Council, he was responsible for making certain that an independent examination of this report was carried out in accordance with institutional procedures and that all review comments were carefully considered. Responsibility for the final content of this report rests entirely with the authoring committee and the institution.

Contents

Executive Summary

Since 1968 the Elementary and Secondary School Civil Rights Compliance Report (known as the E&S survey) has been used to gather information about possible disparities in access to learning opportunities and violations of students' civil rights. Thirty-five years after the initiation of the E&S survey, large disparities remain both in educational outcomes and in access to learning opportunities and resources. These disparities may reflect violations of students' civil rights, the failure of education policies and practices to provide students from all backgrounds with a similar educational experience, or both. They may also reflect the failure of schools to fully compensate for disparities already present as children first enter school that may be due in part to past discriminatory practices and current differences in parents' education, income, and family structure.

The Committee on Improving Measures of Access to Equal Educational Opportunity concludes that the E&S survey continues to play an essential role in documenting these disparities and in providing information that is useful both in guiding efforts to protect students' civil rights and for informing educational policy and practice. There is no other source for much of the information on the E&S survey. The committee also concludes that the survey's usefulness and access to the survey data could be improved.

BACKGROUND

Initially, the E&S survey exclusively addressed questions related to the racial composition of schools. For many years, the survey was the only source of information about school segregation in the United States. Under the authority of Title VI of the Civil Rights Act of

1964, the Office for Civil Rights (OCR) of the U.S. Department of Education used this information to secure the compliance of local education agencies with school desegregation orders. Over the years, items were added to the E&S survey concerning other possible violations of students' civil rights under Title VI, which prohibits discrimination on the grounds of race, color, national origin, and language. Additional questions also have been added to the survey concerning two other laws for which OCR has enforcement responsibility and that were closely modeled after Title VI: Title IX of the Education Amendments of 1972, which prohibits discrimination on the basis of sex in programs that receive federal financial assistance, and Section 504 of the Rehabilitation Act of 1973, which prohibits discrimination against persons with disabilities. Violation of any of these laws can result in the withholding of federal financial assistance, although this penalty rarely has been imposed.

Because the rationale for administering the E&S survey has been to provide information related to the enforcement of these three laws, much more emphasis historically has been placed on the *collection* of data from schools and districts regarding their compliance with these laws than on routine *analyses* and *dissemination* of findings. Information from the E&S survey usually represents a first but important step in the process of determining whether minority students, students with disabilities, students with limited English proficiency, and females are affected by policies and practices that limit access to learning opportunities or resources. By itself, the E&S survey can only be used to identify statistical relationships and disparities in learning opportunities; it cannot be used to address questions of causality. However, statistical associations among certain variables can suggest that there may be unintended negative consequences of educational policies and practices or possible violations of law.

OCR reports that the survey has long been underutilized. Although some OCR enforcement staff reportedly use the data, most do not. No training is provided to OCR staff regarding technical issues in accessing and analyzing E&S survey data or potential uses of the data in enforcement. The data sometimes are used in conjunction with citizen complaints and other information to decide whether potential problems in specific schools and school districts may require further investigation.

Except for a brief period during the 1970s, data from the survey have never been routinely published or otherwise made widely available to the public. OCR has made the data available to education advocates, researchers, and other interested parties on request. However, sometimes this has required use of the Freedom of Information Act by those seeking the information. In the summer of 2002, OCR took a major step to make the E&S survey data much more available to the public by placing them on the Internet. The committee commends OCR for this step.

FINDINGS

For many issues, the E&S survey is the sole source of national data that are disaggregated by racial, ethnic, and language minority status, gender, and by students' disability status. The survey is the sole source of information at the school and classroom level for most of the policies and practices it covers. No other source of information provides a comprehensive national picture of disparities in disciplinary practices in schools, gender disparities in interscholastic athletics, classroom-level student placements in early and middle grades, the effects of high-stakes testing, and services for students with disabilities and limited proficiency with English. Although various national surveys touch on some of the issues dealt with in the E&S survey, no other data source provides disaggregated information that is updated regularly and is traceable to individual class-

rooms, schools, and districts.

With some modifications and closer coordination with other Department of Education databases to accommodate more sophisticated research designs, E&S survey data could play a more prominent role than they have in the past in informing contemporary policy and research questions—especially the identification of possible unintended inequitable outcomes of various educational policies and practices.

The Department of Education has announced its intention to consolidate the collection and maintenance of administrative data used for program management and policy decisions through the proposed Performance-Based Data Management Indicators (PBDMI) system. The implementation of the PDBMI system may affect the way in which data that currently are gathered through the E&S survey are collected in future years. The committee advocates that the department continue collecting data on access to learning opportunities that will help ensure that students with special needs are well served. Data currently obtained from the E&S survey should continue to play an important role in protecting the rights of students from all backgrounds. The fact that the survey provides up-to-date district, school, and classroom-level data distinguishes it as a uniquely important information resource. If the E&S survey were to be incorporated into another data collection instrument, it is essential that the legislatively based requirements mandating the timely provision of specific data at the classroom and school levels be retained.

The E&S survey historically has focused on the enforcement of civil rights issues and complaints. However, it has considerable potential to be used as a complement to the No Child Left Behind Act of 2001 by providing information on educational policies and practices that affect students' access to and the consequences of essential learning opportunities. To do this, the survey should be strengthened in several ways. First, the survey should be integrally linked to other Department of Education data collection efforts, including special education data surveys and the department's proposed PBDMI system. The survey's unique contribution of providing school district, school, and classroom data should be maintained, and its capability to measure trends over time should be expanded.

Second, revising questions and eliminating unnecessary items could improve the E&S survey itself. Field testing, respondent validity studies, and findings from ongoing research on learning opportunities should guide the revisions of the survey. Some examples of items that need revision are provided in the report, e.g., ability grouping and teacher qualification indicators.

Though the survey data have been useful to a variety of users, the datafiles have been difficult to access and use. OCR has placed some of the information on its website, facilitating simple queries of the data. Problems remain for researchers and others who try to do more detailed studies. The E&S survey data should be formatted in ways that make them easier for researchers to use. Improving the ability to connect the data to existing datasets could help with this issue. Also, multiple sources of training and support to OCR staff and other users of the survey data should be provided, and efforts to disseminate survey findings expanded.

CONCLUSIONS

Disparities in educational outcomes and in learning opportunity among different types of students continue to be an important social problem. In that context, the committee has three principal conclusions about the E&S survey:

- The E&S survey, or an equivalent research instrument, continues to be needed to gather disaggregated data related to the equality of access to learning opportunities and resources that are traceable to individual classrooms, schools, and districts.

- The survey, although useful for civil rights enforcement, informing educational policy, and the conduct of research, is greatly underused.

- The survey can be made more useful by improvements to the content, the manner in which the survey is administered, and access to the valuable data it provides.

RECOMMENDATIONS

The committee offers recommendations in four categories: survey administration, improving data quality, increasing access to the data, and disseminating survey findings. We end with an overall conclusion about the role of the E&S survey.

Survey Administration

- The mandated and certified collection of data related to possible violations of students' educational civil rights should be sustained.

- The survey should be supported by line-item funding in the department's budget to ensure its ongoing support at a level that is consistent with its continued quality.

- Because of the survey's importance, the department should consider undertaking a thorough study of the survey aimed at ensuring that it deals appropriately and in sufficient depth with the problems of discovering possible restrictions on students' learning opportunities and, if possible, reducing the reporting burden on schools and school systems.

- The E&S survey content and protocols should be coordinated with those of other department surveys to ensure consistency of definitions and the complementarity of the data and to eliminate redundant questions.

- The various stakeholders in the E&S survey—such as OCR enforcement staff, student advocates, state and local educators, and researchers—should discuss and explore the advantages and disadvantages of less frequent but more comprehensive surveys. With respect to a comprehensive survey, all schools should be surveyed, at a minimum, every 10 years, as was done in 2000.

Improving Data Quality

- Survey items should be revised to provide more useful and complete information on five topics:
  1. the qualifications and experience of teachers;
  2. the assignment of students to different types of classrooms and educational settings;
  3. the consequences for students of high-stakes testing;
  4. high school completion; and
  5. interscholastic athletics.
- OCR should ensure that respondents understand how to complete the survey accurately and thoroughly.
- OCR should carefully scrutinize the data that are collected for thoroughness and reliability.

Increasing Access to the Data

There are several steps that OCR should take to increase access to the E&S survey data:

- train staff to make more effective use of the survey data;
- continue to improve the software provided for public access to E&S survey data over the Internet;
- sponsor or support programs to train advocates, researchers, and educators to use the data for various purposes;
- make well-edited data available to researchers and others in a usable format and provide a data manual and technical assistance;
- consider developing a small grants program to encourage research on the topic of access to learning opportunities using E&S survey data; and
- archive and preserve data from all surveys in a common format and make them accessible to researchers and other interested parties on disk or over the Internet, both for historical purposes and to enable researchers to track longitudinal trends.

Disseminating Survey Findings

Three steps should be taken by OCR to improve dissemination of E&S survey data:

- conduct or sponsor the conduct of basic tabulations of the data;
- include findings from analyses of the data in OCR's regular reports to Congress; and
- publicize the basic findings from the survey in widely disseminated government publications.

1

Introduction

Since the mid-1990s, local and state education agencies increasingly have emphasized the need to hold students, educators, and schools accountable for meeting high academic standards. With the passage of the No Child Left Behind Act in 2001, the need for accountability was confirmed again in federal law, as was the federal government's commitment to ensuring equitable access to essential learning opportunities.

Historically, the Office for Civil Rights (OCR) of the U.S. Department of Education is the federal agency that has been charged with ensuring that all students—especially those in protected classes, as defined in civil rights laws by race, ethnicity, gender, religion, national origin, language, and disability (see below)—have equitable access to learning opportunities. In addition to monitoring complaints from students, parents, and other interested parties, the principal instrument that OCR has

used to identify possible inadequacy of learning opportunities and resources is the Elementary and Secondary School Civil Rights Compliance Report, commonly known as the E&S survey.

The E&S survey was first administered in 1968.[1] In the 35 years since then, profound changes have taken place in the nation and in its schools. For example, in 1968, only 16 percent of the U.S. population consisted of racial or ethnic minorities; by 2000, the percentage of minorities in the U.S. population had nearly doubled to 31 percent (U.S. Census Bureau, 2000), and nearly 4 in 10 school-age children were racial or ethnic minorities (National Center for Education Statistics, 2001). Not only has the minority population grown but it

---

[1]OCR administered a less elaborate survey that was the precursor to the E&S survey in 1967 (see Orfield, 1969).

has become much more diverse and geographically dispersed. In the 1960s, the struggle for civil rights primarily was focused on segregation and other forms of discrimination against blacks—particularly in the South. By 2000, nearly 60 percent of the country's minority population was Hispanic, Asian, and American Indian. Concerns about students' civil rights had also expanded to include a broader range of issues, including equal learning opportunities for students with limited English capabilities, female students, and students with disabilities.

This report examines the continued relevance and adequacy of the E&S survey as a tool for enforcement of civil rights laws in education, for monitoring equality of access to learning opportunities, and for research of other current issues of educational policy and practice. It provides recommendations on how the survey's design, data collection, and analysis can be improved to enhance the survey's value.

THE COMMITTEE AND ITS WORK

In 2002, the OCR, with the support of the Office of English Language Acquisition, asked the National Academies to undertake a study to examine how the E&S survey could more effectively measure student access to learning opportunities and how the resulting data might be made more accessible and useful both to those concerned with the protection of students' civil rights and to the conduct of research.

To this end, the National Academies' Center for Education and Committee on National Statistics collaborated in the formation of the Committee on Improving Measures of Access to Equal Educational Opportunity to study the E&S survey and its uses. The committee's charge was to

- oversee an evaluation of the E&S survey to determine whether it can be used to
  — identify significant trends in the area of access to equal educational opportunity for all students and
  — inform the work of the OCR and the Department of Justice;
- commission papers analyzing several issues covered by the survey, data permitting;
- comment on how the E&S survey methodology can be improved and/or augmented to provide new data and enhance its analytical and evaluative potential; and
- identify ways in which the E&S data can be linked with other datasets to provide a fuller context for analyzing issues related to

access to equal educational opportunity.

In order to address these issues, the committee met three times and commissioned five papers that are based on E&S survey data. The committee's primary objective in commissioning these papers was to learn about the adequacy of the data for purposes of research, and the papers provide part of the evidentiary foundation for this report. Several committee members collaborated in the research and writing of these papers.

Because of the short amount of time available, the committee decided that it would rely on the commissioned papers for the preliminary analyses of issues. The committee encouraged the authors to continue to pursue their research independently and, if warranted, separately publish more fully developed analyses. Synopses of the commissioned papers are provided in Appendix A.

The committee also oversaw a basic analysis of data from the 2000 E&S survey. The 2000 survey is particularly noteworthy because, for only the second time in its 35-year history, it included data from virtually all public elementary and secondary schools in the United States, rather than from a sample of them. Selected findings from this analysis are presented in Appendix B; highlights of this material also are presented in Chapter 2.

In addition to commissioning papers and overseeing a basic analysis of the data, the committee carefully examined and discussed the survey and each of the issues it addresses.[2] The committee also heard from officials from the OCR concerning technical matters related to survey design, administration, and analysis, as well as the uses of the survey for purposes of enforcement and to inform department policies and procedures. The committee and its staff reviewed technical documentation for the survey provided by the department as well as available publications that are based on survey findings. The committee also interviewed individuals who have been involved with or are knowledgeable about how the E&S survey has been used at different points in its history.

The committee's 10 members brought to bear on these issues a wealth of experience and expertise representing a variety of perspectives, including researchers in the fields of education, sociology, economics, public policy, and political science; officials

_____

[2]The issues addressed by E&S survey items are reviewed in Chapter 2. The 2000 survey itself, including instructions and definitions, is provided in Appendix C. A more detailed examination of the complementarity of E&S survey items with data available from other Department of Education datasets is provided in Appendix D.

from state and local education agencies; and the leader of a national student advocacy organization.

This report reflects the committee's consensus views on the survey's strengths and weaknesses and on changes that would make the survey a more valuable information resource for civil rights enforcement, education research, and other efforts to ensure equitable access to learning opportunities. It is important to note that there are very few publications that discuss the E&S survey and surprisingly few published analyses of E&S data that have been produced either by the Department of Education or by independent researchers. Given these circumstances, the report is based primarily on the committee's own evaluation of the survey, analysis of interviews we conducted, and our experience with and findings of the papers we commissioned.

In Chapter 2 we review the features of the E&S survey that make it an important information resource for ensuring equal access to high-quality education. In Chapter 3 we discuss how the survey has been used by the OCR, by education advocates, and by education researchers, as well as possible new uses for it. In Chapter 4 we suggest changes that would make the E&S survey a more valuable information resource for OCR's civil rights enforcement efforts, for

parents who seek to ensure that their children's schools provide them with the kinds of educational services to which they are entitled, and for researchers interested in investigating more complex issues than those for which data from the E&S survey traditionally have been employed. Chapter 4 also discusses strategies for increasing the utilization of the survey by the public and by researchers. In Chapter 5 we conclude with a summary of our conclusions, our recommendations, and a discussion of promising new uses of E&S survey data.

CIVIL RIGHTS LAWS AND THE E&S SURVEY

The Survey's Origins: Title VI and School Desegregation

A civil right can be defined as an enforceable right or privilege, which if interfered with by another, gives rise to an action or injury. The framework for defining the civil rights of Americans is laid out in the U.S. Constitution and Bill of Rights, the Thirteenth and Fourteenth Amendments, and in various laws that have been enacted over the years. One of the earliest and most important of these laws for the enforcement of civil rights in schools is the Civil Rights Act of 1964. Title VI of that act became a powerful tool used by the OCR to desegregate schools. The OCR initially developed the E&S survey exclusively to support that objective.

School desegregation has been described as the cornerstone of the modern civil rights movement (Taylor, 1971; Edley, 2002, p. 126), and it was one of a handful of social issues that defined much of the latter half of the 20th century in the United States. Indeed, the period from the late 1950s into the 1970s is increasingly referred to as the civil rights era (DeFrancis, 1998; Gregg and Leinhardt, 2002) and is the subject of many history courses being taught in colleges around the country.

The origins of civil rights enforcement related to education are rooted in the Supreme Court's 1954 decision in *Brown v. Board of Education*. The Supreme Court ruled in that case that segregating students by race in different schools was a violation of the equal protection clause of the Fourteenth Amendment to the U.S. Constitution. Informed by research and testimony on the role of segregation in maintaining the South's Jim Crow system of allocating privilege on the basis of race (Myrdal, 1944; Clark, 1963; Kluger, 1976; see also National Research Council, 1989), the Supreme Court ruled "separate educational facilities are inherently unequal" (347 U.S. 495, 1954).

Despite the forceful language of the *Brown v. Board of Education* ruling, progress in implementing the ruling was slow, and partly in response, Congress passed the Civil Rights Act of 1964 (Orfield, 1969; Halpern, 1995; Ferguson and Mehta, 2002). Title VI of that act provided the OCR and its predecessor agency, the Equal Educational Opportunity Program, with the statutory authority and a powerful administrative tool for school desegregation. OCR's efforts to enforce Title VI played a pivotal role in fostering the substantial progress that was made toward school desegregation in the late 1960s and early 1970s (Orfield, 1969; Rabkin, 1980; Halpern, 1995; Glennon, 2002).

OCR used the E&S survey to provide quantitative documentation of ongoing problems with segregation throughout the country and to monitor the progress of districts under court-ordered desegregation plans and of districts with which OCR had negotiated desegregation consent agreements (U.S. Commission on Civil Rights, 1969; Orfield, 1969). Title VI states: "No person in the United States shall, on the ground of race, color, or national origin, be excluded from participation in, or denied the benefits of, or be subjected to discrimination under any program or activity receiving Federal financial assistance" (PL 88-352 Title VI, Section 601). Title VI provides OCR and other federal agencies with the authority to deny federal funding to state and local government entities, including school districts that discriminate on the basis of race, color, or national origin.[3]

Language specifying OCR's authority under Title VI to collect information (i.e., through the E&S survey) that is needed for the law's implementation appears in the Code of Federal Regulations.[4] Prior to the creation of the E&S survey, the federal government did not systematically collect racial data in education. Many school districts either did not collect the data or refused to make them available. The E&S survey made it possible for the first time to precisely monitor patterns and trends in segregation and, later, other civil rights issues related to race and ethnicity—a capability that is necessary, although not sufficient by itself, to civil rights enforcement (Orfield, 2001a, 2001b).[5]

Federal court rulings on desegregation have always relied on judicial findings of violations that produced illegal segregation. During the 1960s

---

[3]Prior to 1965, even this important regulatory tool would not have been very effective because the federal government provided very little funding to local school districts. This changed, however, in 1965, with the enactment of legislation that created two important federally funded programs—Head Start, the early childhood education program, and Chapter I (Title I) of the 1965 Elementary and Secondary Education Act, which provided federal financial aid to low-income schools (Orfield, 1969, p. 45; Halpern, 1995, pp. 45–46).

[4]Information is collected in the E&S survey pursuant to 34 C.F.R. Section 100.6(b) of the Department of Education regulations implementing Title VI of the Civil Rights Act of 1964. "Each recipient shall keep such records and submit to the responsible Department official or his designee timely, complete and accurate compliance reports at such times, and in such form and containing such information, as the responsible Department official or his designee may determine to be necessary to enable him to ascertain whether the recipient has complied or is complying with this part. For example, recipients should have available for the Department racial and ethnic data showing the extent to which members of minority groups are beneficiaries of and participants in federally-assisted programs. . . .

Each recipient shall permit access by the responsible Department official or his designee during normal business hours to such of its books, records, accounts and other sources of information, and its facilities as it may be pertinent to ascertain compliance with this part."

Requirements also are incorporated by reference in department regulations implementing Title IX of the Education Amendments of 1972, and Section 504 of the Rehabilitation Act of 1973.

[5] Title VI also led to the creation of an OCR-administered survey on the racial and ethnic composition of colleges and universities. Responsibility for the collection of these data later was transferred to the National Center for Education Statistics' Integrated Postsecondary Education System and its fall enrollment and completion surveys (see http://nces.ed.gov/Ipeds). Also, responsibility for the collection of data on the racial and ethnic composition of elementary and secondary teachers and administrators that were initially collected through the E&S survey later was transferred to the Equal Employment Opportunity Commission, which collects this information at the district level through its EEO-5 survey (see (http://www.mimdms.com/EEO5.html).

and early 1970s, violations were very easy to show in cases that addressed school segregation in 17 southern and border states where segregation was required or authorized by law. Elsewhere, with some exceptions, school segregation was determined to be "de facto"—i.e., the product of segregated neighborhoods (Orfield, 1969, 1978). However, the Supreme Court ruled in the 1973 *Keyes* case that, in school settings outside the South, an entire school district could be presumed to be illegally segregated once the plaintiffs had proved that there were acts of intentional discrimination that affected a significant portion of the district. The *Keyes* decision extended remedies both to school districts outside the South and to Hispanics, whose segregation had not been addressed in *Brown v. Board* and related decisions (Orfield, 1978).

Only a year later, progress in desegregating schools was slowed when a closely divided Supreme Court decided in the 1974 *Milliken v. Bradley* case that urban desegregation plans could not include suburban districts except in very exceptional circumstances. The Court's ruling in *Milliken v. Bradley* substantially ended major desegregation efforts in many of the largest metropolitan areas, except those with countywide districts that primarily were located in the South and the West, such as Louisville, Kentucky (Orfield, 1996; Rebell,

2002). Then, in 1977, Congress passed an amendment to the 1964 Civil Rights Act that prohibited OCR from requiring school districts to use busing as a strategy to desegregate schools; this further reduced OCR's ability to carry out its original mandate to desegregate schools by enforcing Title VI (Halpern, 1995, pp. 154–160). Finally, the 1981 termination of the Emergency School Aid Act, which was used to implement the provisions of the federal desegregation aid program, further curtailed OCR's effectiveness in promoting desegregation (Orfield, 2000).

Together, these events greatly diminished OCR's role in promoting school desegregation. Nevertheless, OCR has continued to use the E&S survey to monitor the racial and ethnic composition of schools in hundreds of districts with which it previously had entered into settlements as well as in other districts that have remained under court-ordered desegregation plans.

Other Title VI Enforcement Issues

Although desegregation is no longer a focus of OCR civil rights enforcement activities, the E&S survey continues to play a role in OCR's enforcement of Title VI. With information routinely gathered from the E&S survey, OCR-initiated compliance reviews, and complaints submitted by parents and other concerned parties, OCR monitors for and acts on a wide variety of possible violations of civil rights under Title VI (U.S. Department of Education, Office for Civil Rights, 1999), including:

- disciplinary policies and practices,
- ability grouping,
- access to language services by English-language learners,[6]
- interdistrict student transfers,
- student assignment policies, including to gifted and talented programs,
- racial harassment, and
- academic grading.

Other Civil Rights Laws and the Evolution of the E&S Survey

Section 504

In addition to Title VI, OCR also is responsible for ensuring equal educational opportunity and protecting students' civil rights based on other laws. Chief among these is Section 504 of the Rehabilitation Act of 1973[7] that bars discrimination on the basis of disability. With the support of information derived from the E&S survey, OCR addresses the following issues pertaining to students with disabilities:

- teaching students in the least restrictive environment consistent with their educational needs,
- suspensions and expulsion of students with disabilities,
- appropriate special education services,
- academic adjustments and modifications, and
- auxiliary aids for students with impaired sensory, manual, or speaking skills.

Title IX

OCR also monitors and addresses issues related to gender equity under the authority of Title IX of the Education Amendments of 1972,[8] including:

---

[6]After the Supreme Court's ruling in *Lau v. Nichols* in 1973, ensuring the education civil rights of language minority students was added to OCR's Title VI enforcement responsibilities.

[7]Section 504 of the Rehabilitation Act of 1973 requires that "no otherwise qualified individual with a disability in the United States . . . shall, solely by reason of her or his disability, be excluded from the participation in, be denied the benefits of, or be subjected to discrimination under any program or activity receiving Federal financial assistance" (see http://www.ed.gov/offices/OCR/docs/placpub.html).

[8]Title IX states: "No person in the United States shall, on the basis of sex, be excluded from participation in, be denied the benefits of, or be subjected to discrimination under any program or activity receiving Federal financial assistance" (see http://www.ed.gov/offices/OCR/dox/tix_dis.html).

- equal opportunity in interscholastic sports,
- treatment of students who are pregnant, and
- access to and placement in various school programs.

See Appendix C for the specific items on the E&S survey that address issues covered by Title VI, Section 504, and Title IX.

OCR AND EDUCATION AS A CIVIL RIGHT IN THE 21ST CENTURY

The mission of the Office for Civil Rights is to ensure equal access to a high quality education for all students through the vigorous enforcement of civil rights. (U.S. Department of Education, Office for Civil Rights, 2000a, p. 1)

Current efforts to ensure equal access to learning opportunities and resources are not as visible to the public as efforts to end segregation in the civil rights era. Nevertheless, equal access to a high-quality education remains an essential civil right under both federal and state laws (Rebell, 2002).[9]

OCR received 4,897 civil rights complaints in 2000. Advocates for students' civil rights report that persons filing complaints sometimes use E&S data when filing complaints. A majority, 55 percent, involved students with disabilities under Section 504 (U.S. Department of Education, Office for Civil Rights, 2000a); 18 percent of complaints (870) alleged discrimination on the basis of race or national origin (Title VI); allegations of sex discrimination under Title IX accounted for 8 percent (396) of complaints; and 11 percent (539) of complaints alleged discrimination under multiple laws, such as the inappropriate assignment of minority students to special education (both Title VI and Section 504). In 2000, OCR reports that 2,000 school districts and institutions of higher education changed their policies, procedures, or practices to comply with federal civil rights laws as a result of OCR intervention (U.S. Department of Education, Office for Civil Rights, 2000a). Data from the E&S survey also can be used by parents and other citizens, who may pursue grievances against their local schools independently of OCR.

---

[9]The mission statement of the Office for Civil Rights refers to ensuring "equal access to a high quality education" through the enforcement of civil rights laws such as Title VI (U.S. Department of Education, Office for Civil Rights, 2000a). In addition, nearly all state constitutions

cite the state's responsibility to establish a system of "free and common schools" to provide students with "a thorough and efficient education," "an adequate public education," or an "ample education" (Rebell, 2002, p. 232).

These data concerning OCR enforcement activities, as well as information presented in Chapter 2 and in Appendixes A and B of this report, suggest that violations of students' civil rights are not uncommon and that major disparities in access to various kinds of learning opportunities remain. The E&S survey is the only vehicle now available to identify many of these disparities. OCR reports that it uses the survey data, in conjunction with citizen complaints and other information, to decide whether potential problems in specific schools and school districts may require further investigation. As is discussed below, others also use E&S survey data to identify disparities in access to learning opportunities and resources.

The consequences for students' inability to have access to a high-quality education may be as serious as they were during the civil rights era because of the dwindling number of well-paying jobs requiring little education (see Moses and Cobb, 2001). For example, as recently as 1979, male college graduates earned only 29 percent more than male high school graduates and 57 percent more than male dropouts. By 1999, the earnings advantage of male college graduates over their high school graduate and dropout counterparts had increased to 68 percent and 147 percent, respectively. The real wages of male

workers with only a high school education or less have fallen steadily in the past 20 years (Council of Economic Advisers, 2000, pp. 135–136).[10]

Students attending predominantly minority, high-poverty schools are particularly at risk of not getting the kind of education that will prepare them for the 21st century workforce. These students, on average, are much less likely than others to graduate from high school (Neild and Balfanz, 2001; Roderick and Engel, 2001), and they have substantially lower test scores (Lippman et al., 1996; Puma et al., 1997; Balfanz et al., 2002). Based in part on an analysis of data from the E&S survey, Orfield (2001b) found that in schools in which 50–60 percent of the students are black or Hispanic, on average, at least 60 percent of the student population are poor. In schools in which at least 80 percent of the students are black or Hispanic, on average, 80–90 percent of the students are poor (Orfield, 2001b). Although it is methodologically difficult to differentiate the effects on learning outcomes of poverty and other factors associated with students' neighbor-

---

[10]Real wages of female high school graduates and dropouts have dropped much less than those for their male counterparts because such women historically have been much less likely to hold well-paying jobs (see National Research Council, 2002a, p. 16).

hoods and families from those of schools (see Halpern-Felsher et al., 1997), data from the E&S survey and other sources (see Darling-Hammond, 1997; Adelman, 1999; National Research Council, 1999a; Farkas, 2002) provide evidence that students in high-poverty, segregated schools have less access to learning opportunities and resources than other students (see also Chapter 2 and Appendix B).

Some of the resources and learning opportunities to which students are guaranteed are stipulated by law (e.g., educating students with disabilities in the least restrictive environment consistent with their needs). For most issues, however, the courts and education civil rights laws provide more general guidance on the meaning of equal access to learning opportunities and resources. The emergence of new educational policies and practices and ongoing research on their educational effects and possible civil rights implications periodically have led to the appearance and disappearance of various items on the E&S survey.

The E&S survey produces important information about disparities among students of different backgrounds that may suggest unequal access to learning opportunities and resources. Data from the E&S survey can be used to help identify schools that may be denying students equal access to educational opportunities and resources, as defined by the civil rights laws on which the survey is based. School administrators are required by law to maintain and accurately provide the information elicited by the survey to support the enforcement of the civil rights laws.

By themselves, statistical disparities associated with race, ethnicity, language, gender, and disability status identified through the E&S survey, such as those presented in Chapter 2 and Appendix B, do not necessarily prove that discrimination has occurred or that students' civil rights have been violated. Statistical associations among variables on the E&S survey can reflect possible violations of the law that should be addressed by OCR, the unintended negative consequences of educational policies and practices, the indirect effects of discrimination in housing or employment, other forms of social inequality, or the combined effects of all of the above. Data from the E&S survey could be linked with other datasets in ways that could address questions of causality (see Appendixes A and D). However, the hierarchical structure of the survey itself is a powerful resource for pinpointing the specific schools and districts where problems may be occurring. The fact that E&S data are required to be submitted from such a large sample of schools, many of which are surveyed every two years, can provide a degree of specificity to the analysis of salient issues that would not otherwise be possible from voluntary surveys administered to a much smaller sample of schools.

Whatever the cause or causes, evidence of disparities among groups of students is information that is useful not only for civil rights enforcement, but also for informing efforts to achieve the twin goals of educational excellence and equity, as articulated in the No Child Left Behind Act. The E&S survey's capacity to provide disaggregated information about the distribution of various kinds of disparities for categories of students identified in civil rights laws, and to do so for specific, identifiable schools and districts, is a useful starting point for more comprehensive analyses of the underlying causes of inequality. This information is important not only for civil rights enforcement by OCR, but also for use by parents and other concerned citizens with civil rights concerns or who seek to improve educational opportunities and outcomes for children.

2

Measuring Equal Opportunity:
The Role of the E&S Survey

In education, civil rights always have been about equal access to the opportunity to learn for students from all backgrounds. In the 1960s, this was manifested primarily in the struggle to desegregate schools—and the sole purpose of the Elementary and Secondary Civil Rights Compliance Report was to provide information in support of that goal. In the greatly transformed educational landscape of the first decade of the 21st century, it is important to envision specifically what it means to ensure equal access to a high-quality education.

Over the years, the E&S survey has broadened its focus to include items that address a wide variety of potential violations of students' civil rights. It has become an important, albeit underutilized, source of information regarding the prevalence of educational policies and practices that can restrict students' learning opportunities. The Office for Civil Rights (OCR) of the U.S.

Department of Education is obligated not only to take action against school districts that *intentionally* discriminate against students based on race, color, or national origin, but also Title VI regulations[1] give OCR the authority to take enforcement action against educational policies and practices that result in discrimination (unjustified disparities).[2] The regulations prohibit recipients of federal financial assistance from using "criteria or methods of administration which have *the effect* of subjecting

---

[1] Section 602 of Title VI of the 1964 Civil Rights Act requires federal agencies to create regulations to implement the law (Ryan, 2002). The regulations specifying the Title VI enforcement responsibilities of the Office for Civil Rights are codified in Title 34, Subtitle B, Chapter 1, Part 100 of the Code of Federal Regulations.

[2] The disparate impact of a contested practice is unjustified if it can be shown that the disparity is caused by the contested practice and that the practice either does not serve the legitimate educational goals of the institution or effective alternative practices are available (Ryan, 2002). See Chapter 3 for a more complete discussion of this topic.

individuals to discrimination because of their race, color, or national origin." Policies and practices that result in discrimination may be barred through enforcement action by the department (Losen and Welner, 2002; Ryan, 2002).

It is important to note that the E&S survey was not designed primarily for use by academic researchers. Historically, its purpose has been to provide information to OCR and to members of the public, upon request, related to the compliance of individual schools and school districts with civil rights laws. As noted in an evaluation of the survey prepared for the Department of Education (WESTAT, 1997, p. 1), "OCR conducts the E&S survey to provide its regional offices with current data to use when targeting compliance review sites or to use as source material when investigating complaints." However, the E&S survey is also used by civil rights advocacy groups for monitoring issues related to their mission and to inform and mobilize communities in school improvement efforts. Finally, the survey is used by social scientists conducting research on equality of access to high-quality education. This chapter describes characteristics of the survey that make it useful for civil rights enforcement, for research on patterns of access to learning opportunities, and as an information resource to inform public policy.

SURVEY CHARACTERISTICS

This section describes key features of the E&S survey. The entire 2000 survey is reproduced in Appendix C.

Mandatory and Certified as Accurate

The official name of the E&S survey is the Elementary and Secondary School Civil Rights Compliance Report. Unlike the surveys administered by the Department of Education's National Center for Education Statistics (NCES), for which response is voluntary, schools and school districts are required to respond to the E&S survey and failure to respond could result in the loss of federal financial assistance.[3] Response rates are close to 100 percent, so that OCR has a powerful mechanism for measuring compliance with civil rights laws and other public policy purposes.

The survey has two parts, ED 101 and ED 102. ED 101 is sent to the superintendents of school districts (local education agencies), who are required to certify that the data that they (or their designees) provide about their districts

---

[3]As described in footnote 2 of Chapter 1, this authority is derived from Title VI regulations. It is important to note, however, that since the 1970s, the Office for Civil Rights rarely has used this power as a sanction for violation of civil rights (Halpern, 1995) or for failure to respond to the E&S survey (Rabekoff, 1990).

are complete and accurate. ED 102 is sent to school principals, who are similarly required to certify that the information they provide is complete and accurate. The requirement that the data be certified is meant to deter respondents from reporting false information to OCR. Since data on race were not required in most states before the OCR regulations and some states had policies prohibiting them (Orfield, 1969), it is unlikely that state and local school authorities would collect the key OCR data elements without a mandatory policy.[4]

Data Identifiable for Specific Demographic Groups

The E&S survey contains data on access to opportunities to learn that are broken out by race, ethnicity, gender, and disability status—i.e., for those classes of students whose rights are specified in the civil rights laws that are the basis of the E&S survey. Basic racial and ethnic enrollment data were not routinely collected in the Common Core of Data (CCD) until 1987, 19 years

after the launch of the E&S survey in 1968. Thus, the E&S survey provides the longest continuous record on the racial and ethnic composition of schools, as well as data disaggregated by race and ethnicity on participation in, and the consequences of, various educational practices of any Department of Education survey.

Besides race and ethnicity, the survey has also disaggregated data on a variety of topics by gender, English proficiency, and disability status since the mid-1970s. For many issues addressed by the survey, the data can be disaggregated by multiple categories (e.g., race by gender or race and gender by disability status). Although other Department of Education surveys provide school information on many of the same issues that are covered by the E&S survey, few provide data that are disaggregated to show potential disparities associated with race, ethnicity, gender, language minority status, and disability status.[5]

Sampling and Timeliness

Generally, the Department of Education administers the E&S once every

---

[4]Disaggregated data on opportunities to learn generally are not available in most member countries of the Organisation for Economic Co-operation and Development. For example, in some countries (e.g., France), it is illegal to collect and report data that are disaggregated by race (Orfield, 2001a). In certain states, including California, serious consideration has been given to prohibiting state agencies from compiling race data (see *Sacramento Bee*, May 1, 2002).

[5]Appendix D provides a detailed discussion of the similarities and differences of items on the E&S survey and to those from the CCD, the Early Childhood Longitudinal Study—Kindergarten Cohort (ECLS-K), and the 1988 National Educational Longitudinal Study (NELS:88).

two years, in the even-numbered years. Between 1968 and 1972, the survey was administered annually. The survey was not administered in 1996. Because the survey is administered so frequently, the data about schools generally are more up to date than those available from other Department of Education databases, such as those derived from the NELS-88 or the High School and Beyond surveys. The frequency of the survey makes the data very useful in identifying patterns and changes in school district practices and results.

Except for 1976 and 2000, when the survey was administered to all schools and school districts in the United States, roughly one-third of the approximately 15,000 school districts in the country are included in the survey. School districts with at least 25,000 students are included in each administration of the survey so that sampled districts always include more than one-third of the nation's schools and students. For example, in 1998, 37 percent of all school districts were included in the sample. These districts contained 61 percent of the nation's public elementary and secondary schools and 77 percent of public school students enrolled in grades 1–12.[6]

---

[6]These percentages are calculated from fall 1998 E&S Time Series Documentation (unpublished data, Office for Civil Rights, 2000) and the Digest of Education Statistics (2001, Tables 40 and 89).

Large urban school districts, which have a disproportionate share of minority students and students in poverty, have been included among the sampled districts in every administration of the survey. Also included in each administration of the survey are school districts that are under court order to eliminate civil rights violations. Because most of these districts are in the South, the survey has provided a more comprehensive view of districts in the South than in other parts of the country. Finally, a sample of smaller districts is included in each survey administration (WESTAT, 1997). Overall, the survey provides a continuous record on school civil rights issues that spans more than three decades and is disaggregated to show information for students from groups that are at greater risk of school failure.

Units of Analysis

The units of analysis for the E&S survey are school districts, individual schools, and selected classrooms within elementary schools. The E&S survey is a rare source of information for classroom-level data in the elementary grades. The survey collects information on the racial and ethnic composition of each classroom for the lowest grade in an elementary school (e.g., grade 1) and the highest grade (e.g., grade 6) and whether any ability grouping is used for instruction in those classrooms. However, classroom assignment data are collected only from elementary schools in which the percentage of minority students is more than 20 percent but less than 80 percent.

The survey contains disaggregated group data for all schools in the sampled school districts. Obtaining information from nearly 100 percent of schools in the sampled districts enables one to examine issues for which more selective sampling strategies would not provide a sample of adequate size. Thus, the survey enables examination of disaggregated data at the school and district level that would not be possible using information derived from surveys with more selective sampling techniques.

Although the primary use of the survey has been to identify patterns that suggest potential violations of civil rights in individual schools and districts, the OCR also produces state and national data projections from the sampled schools.

Linking E&S Data to Other Databases

Because the E&S survey is administered so frequently to such a large sample, it is more costly and difficult than it is for smaller scale surveys to include additional questions that would help explain the significance of the simple correlations that can be identified with E&S survey data alone. This is a limitation that the E&S survey shares with other large-scale administrative surveys used for civil rights enforcement—e.g., EEO-1 of the Equal Employment Opportunity Commission and the Home Mortgage Disclosure Act survey of the U.S. Department of Housing and Urban Development (Ross and Yinger, 2002). Without additional information obtained from qualitative research or by linking to data from other surveys, the E&S survey can only highlight patterns of disparity that suggest problems of educational equity. Fortunately, E&S data can be linked to other datasets, considerably expanding their value; the E&S survey includes a school identifier code through which data for individual schools may be linked to information about the same schools from other federal and state surveys.

All of the analyses of E&S data overseen by this committee involved some linkage with the Department of Education's CCD, other databases from the NCES, or education databases maintained by individual states.[7] The

---

[7]In addition, the Census Bureau and the Department of Education are preparing population data from the 2000 census that can be mapped to school attendance boundaries. Population data corresponding to school attendance also will be able to be linked to E&S survey data.

analyses addressed issues related to the following topics:

- student discipline,
- services for English-language learners in first grade,
- the effects of high-stakes testing,
- gender equity in interscholastic sports,
- the characteristics of schools serving large numbers of English-language learners, and
- the grouping of students by race and other characteristics in individual classrooms.

Many analyses overseen by the committee found that linking E&S data to other databases made it possible to investigate research questions that otherwise could not have been addressed (see Appendix A). Because of the short amount of time available, the research conducted under the auspices of the committee necessarily was preliminary and exploratory in nature. The objective was to determine the feasibility of using E&S survey data to investigate various issues, and, if possible, to begin analyses that could lead to papers that could be published independently in peer-reviewed journals. Initial findings from some of these analyses are presented below.

E&S SURVEY AS THE SOLE SOURCE OF NATIONAL INFORMATION

The E&S survey is the only source of national data on school disciplinary practices, gender equity in sports, services for students who become pregnant, and classroom-level data on student assignment.

Classroom-Level Student Placement

The E&S survey is the only national database with information on the placement of all students in classrooms by race and ethnicity and English proficiency.[8] The item also asks whether any students are "ability grouped for instruction in mathematics or English-Reading-Language Arts" in that classroom. Information on classroom assignment is requested for the lowest and highest elementary grades only.

The civil rights concerns emanate from evidence that many students who are "tracked" on a continuing basis into separate classrooms or groups within classrooms because of their below-grade-level performance continue to lose academic ground in these settings,

---

[8]Classroom-level data for elementary schools also are available from the ECLS-K survey for sampled classrooms in the early elementary grades only. Some states also maintain these data.

that minority students are disproportionately "tracked" into low-ability classes, and that such practices may produce "within-school segregation" (see Oakes, 1990; Mickelson, 2001).

Classroom placement data disaggregated by race and ethnicity have been elicited by the E&S survey since 1969, when they were first used to identify what has been called second-generation segregation or within-school segregation.[9]

Disciplinary Practices

In 2000 the E&S survey provided information on corporal punishment, out-of-school suspensions, total expulsions, expulsions that constitute total suspension of educational services, and expulsions due to zero-tolerance policies. The survey also provides information on suspensions and expulsions of students with disabilities. E&S data are disaggregated to show the frequency of various practices by race and ethnicity, for English-language learners, and for students with disabilities. Information

on disciplinary practices has been on the survey since 1973. There is no other national database on school disciplinary practices.

Substantial racial disparities exist in the administration of disciplinary practices. Data from the 2000 E&S survey show that although only 17 percent of all U.S. students were black, 39 percent of the 342,031 students receiving corporal punishment were black, as were 34 percent of the more than 3 million students who received an out-of-school suspension (Appendix B). Of course, out-of-school influences on students' behavior may affect the frequency with which various disciplinary practices are applied for different groups of students. However, these data also suggest one or both of the following: schools are applying different disciplinary standards for students of different races, or, as with other E&S survey indicators, educators are failing to effectively educate, motivate, or engage a large number of black students in the purposes and programs of schools. Any serious effort to ensure equal access to a high-quality education requires that the underlying causes be investigated and addressed. Using E&S survey data to examine differences in the application of disciplinary measures among schools and among districts, as well as longitudinal trends, can be part of this effort.

---

[9]As with other survey items, disparities in the distribution of groups of students in different classrooms may occur for various reasons, and some may be well justified—such as the grouping of English-language learners for language instruction. Although the classroom placement data on the E&S survey are a unique resource, the data could be made much more useful with some minor changes, as is discussed in Chapter 4.

Gender Equity in Sports

The E&S survey is used to gather data pursuant to the enforcement of Title IX of the Education Amendments of 1972. The E&S survey includes information on the number of interscholastic sports teams on which male and female students are eligible to participate. It also includes information on the number of male and female students who participate in interscholastic athletics in each surveyed school. The E&S survey has been used to collect data on gender equity in sports since 1994 and there is no other national database on gender equity in interscholastic athletics.

The 2000 survey showed that, on average, high schools offer 9.7 interscholastic sports teams for boys and 8.7 teams for girls. Of the nearly 6 million high school students who participated in interscholastic sports in 2000, 58 percent were male (see Appendix B).

The Education of Pregnant Students

The E&S survey collects data on educational services for students who are pregnant or who have become mothers. The survey includes items regarding the number of students who became pregnant in the year prior to the survey, and then asks how many of those students (who had not already graduated) were enrolled the following year. This question first appeared in 1980, was dropped from the survey, and then returned in 2000.

E&S DATA THAT COMPLEMENT OTHER DATA

Although similar data on several topics can be obtained from other sources, they are less likely to be disaggregated, to be traceable to specific schools and districts, or to be as current as E&S data. These data cover teacher certification, the consequences of high-stakes testing, the characteristics of special-purpose schools, the number of graduates and type of diploma awarded, segregation, English proficiency, advanced placement classes, gifted and talented programs, and students with disabilities. Most other national survey data tend to be seriously out of date and to have samples that are much too small to be used to study individual states, let alone districts or individual schools.

Teacher Certification

Research indicates that the qualifications of teachers is the school resource that has the greatest impact on student learning outcomes (Hedges, Laine and Greenwald, 1994; Sanders and Horn, 1995; Ferguson and Ladd, 1996; Greenwald, Hedges, and Laine, 1996;

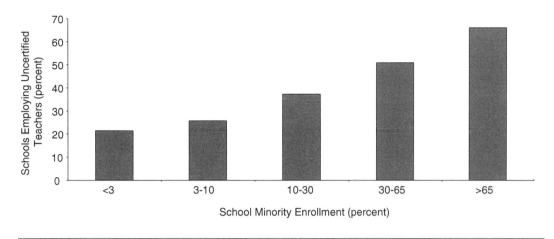

FIGURE 2-1 Minority enrollment and schools employing any uncertified teachers.

SOURCE: Data are from the 2000 E&S survey.

NOTE: Data do not include special education or alternative schools (school n = 82,341). Sixty percent of all public schools employ only state-certified teachers. Minority enrollment includes American Indians, blacks, and Hispanics.

Sanders and Rivers, 1996; Darling-Hammond, 1997a; Ferguson, 2000). The E&S survey contains items about the total number of teachers in a school and how many of the teachers are fully certified. Data from the 2000 E&S survey show that schools with predominantly minority students (black, Hispanic, and American Indian) were significantly more likely to employ uncertified teachers; see Figure 2-1.

High-Stakes Testing

The E&S survey includes information about the number of students who passed or failed district- or state-mandated high-stakes tests, as well as the number who were given alternative assessments or were not tested. The data are disaggregated by race, ethnicity, gender, English proficiency, and for students with disabilities. Data disaggregated in this way provide an important opportunity to investigate the immediate consequences of high-stakes testing and how accountability policies affect the educational opportunities of specific student populations. Data are presented for each of the primary grades in which students are required to pass a test as a criterion for promotion to the next grade. At the secondary level, data in the same format are collected regarding tests that are required as a condition for graduation.

Because of rapid changes in education policies in recent years, the contemporaneousness of the E&S data makes them especially important. Most other recent surveys that contain data about high-stakes testing focus on testing policies and practices being advocated by federal and state legislators and policy makers (see American Federation of Teachers, 2001; Council of Chief State School Officers, 2000; Editorial Projects in Education, 2001, pp. 68–87). However, there is often an important difference between formally stated public policies and the actual implementation of those policies by practitioners in schools and school districts (Cuban, 1998). The E&S survey provides insights into these differences.

That the E&S survey contains data on high-stakes testing, along with information about enrollment changes, special education placements, disciplinary information, and other data, is particularly important in light of the increasing use of high-stakes testing. Education advocates and others are especially concerned that the standards movement and associated high-stakes tests could create incentives for schools to inappropriately place poorly performing students in special education (see National Research Council, 2002b, p.85), transfer poorly performing students to other schools, or take other actions that would eliminate them from the pool of tested students.

Analysis of the 2000 survey conducted for this report revealed that districts that were predominantly black and Hispanic were far more likely to require students to pass a district- or state-sponsored examination as a condition for grade promotion or graduation (Croninger and Douglas, 2002). Also, of those students required to take high-stakes exams, black and Hispanic students are more likely than white and Asian American students to have failed; see Figure 2-2.

Characteristics of Special-Purpose Schools

The E&S survey is an important source of information on the student composition, aspects of the curricula, and certain educational resources and practices of specialized schools. The survey (ED 102) asks whether the reporting school is one of five kinds:

- a magnet school,
- a charter school,
- an alternative school for students with academic difficulties,
- an alternative school for pregnant students, or
- an alternative school for students with discipline problems.

Information on the number of charter schools, magnet schools, and alternative schools is also collected in the CCD,

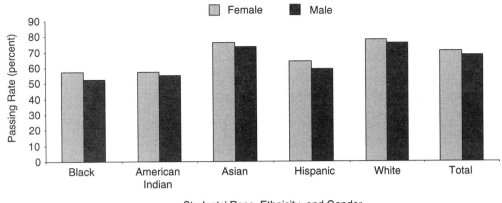

FIGURE 2-2 Passing rates for tests that are the sole criterion for high school graduation.
SOURCE: Data are from the 2000 E&S survey.
NOTE: Schools were instructed to consider a test of the "sole criterion" for graduation if "all students were required to take a district- or state-required test, and must pass the test to graduate from high school." A total of 2,652 high schools required these tests; 585,709 high school students took such tests; 406,502 passed.

and in 2002 the NCES (Kleiner, Porch, and Farris, 2002) published a statistical analysis on public alternative schools and programs for students at risk of education failure. Little is known about how some of these schools—especially charter schools—are similar to or different from other schools. According to E&S survey data for 2000, minorities account for 40 percent of the enrollment of charter schools, compared with 30 percent of regular public schools. Charter schools are much more likely to employ teachers who are not fully certified.

Number of Graduates and Type of Diploma Awarded

Schools are required to report the number of students awarded regular diplomas and certificates of attendance or certificates of completion. Data from the 2000 E&S survey showed that blacks and American Indians were substantially more likely than whites and Asian Americans to a receive a certificate of attendance or completion instead of a diploma; see Figure 2-3.

Segregation

As discussed above, the original purpose of the E&S survey was to

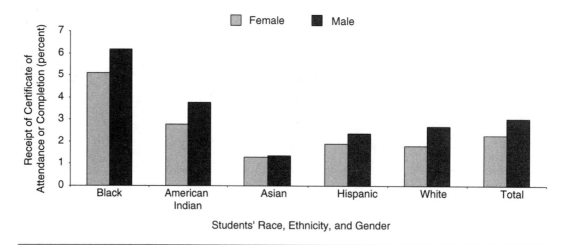

FIGURE 2-3 Certificates of attendance or completion to high school graduates.
SOURCE: Data are from the 2000 E&S survey.
NOTE: OCR projections indicate 2,605,843 public high school graduates in 2000, of whom 69,081 received certificates of attendance or completion. OCR defines a certificate of attendance or completion as "an award of less than a regular diploma, or a modified diploma, or fulfillment of an IEP for students with disabilities."

document patterns of racial segregation and the compliance of districts with desegregation plans. Since 1987, information on the racial and ethnic composition of schools comparable to that collected on the E&S survey has been collected annually in the CCD. However, the E&S survey is the only source of information for earlier years, and it has proven to be an indispensable resource for documenting long-term trends, including the recent trend toward resegregation (see Orfield, Bachmeier, and Eitle, 1997; Orfield, 2001a). Data from the 2000 survey, when combined with information on school poverty from the CCD, show that highly segregated black and Hispanic schools also tend to have a very high concentration of students in poverty; see Figure 2-4.

Because nearly all E&S survey items are disaggregated by race and ethnicity, the survey is an important source of information on what has been called second-generation segregation or within-school segregation—an issue of particular interest for access to high-level classes and curriculum tracking.

English-Language Learners

The E&S survey also is an important source of information concerning English-language learners and the

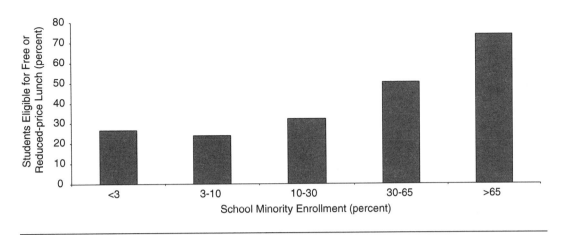

FIGURE 2-4 School minority enrollment and poverty concentration.
SOURCES: Data are from the 2000 E&S survey (minority enrollment) and the 2000–2001 Common Core of Data (free and reduced-price lunch eligibility).
NOTE: The data do not include special education and alternative schools (school n = 69,029). Minority enrollment includes American Indians, blacks, and Hispanics. Data on free and reduced-price lunch eligibility are not available for Arizona, Connecticut, Illinois, Tennessee, or Washington.

schools that serve them. According to 2000 survey data, 3.2 million (8 percent) of elementary and secondary students need English-language learner services: 76 percent of these students are Hispanic, and 13 percent are Asian American. According to the 2000 E&S survey, 36 percent of all Hispanic students need English-language learner services, as do 24 percent of all Asian American students.

Hispanic English-language learners, in particular, tend to be concentrated in highly segregated, high-poverty schools (Horn, 2002). On average, 60 percent of students in schools attended by His-panic English-language learners are from families in poverty, and only 22 percent are non-Hispanic whites. Nearly half (45 percent) of all Hispanic English-language learners attend schools in which at least 90 percent of the students are minorities (see Appendix B). The isolation of students in segregated, high-poverty schools tends to concentrate disadvantage and can contribute to poor educational outcomes (Lippman et al., 1996; Puma et al., 1997). The E&S survey is an important source of data that can be used to analyze this problem. In addition, the survey is a useful source of information on the availability of learning resources in the schools attended by English-language learners and on their access to those resources within their schools (Horn, 2002).

An official from the Office of English Language Acquisition informed the committee that the E&S survey is particularly valuable for its purposes because the data are current. The number of English-language learners is growing rapidly, and the E&S survey documents how they have become more widely dispersed across many more states. The survey also identifies the districts and schools where these students are concentrated.

Advanced Placement Classes

The E&S survey provides information on the number of advanced placement (AP) courses taught in high schools. On average, schools with predominantly minority students offer fewer AP courses than schools with predominantly white students (see Appendix B).

The survey also provides information on the number of students taking AP science and mathematics courses, disaggregated by gender, race, and ethnicity. In 2000, black, Hispanic, and American Indian students were about half as likely as white students to be enrolled in AP science and math courses (see Appendix B). The E&S survey also is a source of information on the number of students who have limited English proficiency and learning disabilities who are enrolled in AP classes. Information on access to AP courses has been available from the E&S survey since 1992.

Gifted and Talented Programs

The E&S survey provides information on the characteristics of elementary school students who are enrolled in gifted and talented programs. The 2000 survey showed that black, Hispanic, and American Indian students were 40 percent, 49 percent, and 64 percent, respectively, as likely as white students to participate in gifted and talented programs (see Appendix B).

Students with Disabilities

Since 1976, the E&S survey has included information on the composition of students in various disability categories and the degree to which those students are placed in inclusive educational settings—that is, the extent to which they are either "mainstreamed" or served in separate classrooms or schools.

This information is gathered because of OCR's responsibility to ensure compliance with Section 504 of the Rehabilitation Act of 1973, which prohibits discrimination based on disability. As with other items, the E&S survey disaggregates information on students with disabilities by race and ethnicity, and, in some cases, gender. Although surveys administered by the Office of Special Education and Rehabilitative Services (OSERS)—previously called the Office of Special Education Programs (OSEP)—have long collected state-level data on students in special education, the OSEP/OSERS data were not disaggregated by race and ethnicity until 1998 and have never been reported for individual schools and districts.

Prior to 1998, the E&S survey was the only source of routinely collected information on the racial, ethnic, and language background of students enrolled in special education. Recently, OSERS began to collect this information as part of its regular survey conducted for the purpose of monitoring states' administration of the Individuals With Disabilities Act, as amended in 1997. That year, Congress required OSEP to begin monitoring racial and ethnic disparities in special education placement for the possibility of inappropriate placement (see Hehir, 2002).

Since OSEP began to collect disaggregated data in 1998, the data collected by OSEP/OSERS have become essentially redundant with the items on the E&S survey regarding students with disabilities. In the interest of eliminating redundancy and reducing the paperwork burden on respondents, OCR and OSEP/OSERS pilot-tested a combined

data reporting instrument in 2000. The combined survey instrument covering items required by both OCR and OSEP/OSERS on students with disabilities was administered by OSEP/OSERS.[10]

Racial disparities in certain categories of special education have long been a matter of concern (see National Research Council, 1982). Part of the concern has centered on the possibility that minority students were being inappropriately evaluated for placement in special education. In addition, there is concern that the poor quality of regular education services available to many minority students in the early grades may result in the disproportionate placement of students from certain minority groups in special education as they reach third grade (Hehir, 2002; National Research Council, 2002b). Furthermore, E&S survey data show that black and Hispanic students, once they had been identified as needing special education services, are far less likely to be educated in a fully inclusive general education classroom (Garcia Fierros and Conroy, 2002).

---

[10]The Department of Education is in the process of developing a much more comprehensive approach to the collection of administrative data from schools and school districts, known as the Program-Based Data Management Initiative (PBDMI). The possible implications of PBDMI for the E&S survey are discussed in Chapter 3.

An analysis of E&S survey data for a National Research Council report (Finn, 1982) found that some minority groups—especially blacks—were disproportionately represented among the "educable mentally retarded" and among students labeled as "emotionally disturbed." The report also found that males were significantly overrepresented in these and other categories of special education. Using 1998 E&S survey data, a later National Research Council report (2002b) documented essentially similar disparities.[11]

Another recent report on the topic, *Racial Inequity in Special Education* (Losen and Orfield, 2002b), includes a number of papers that also used E&S survey data to document racial disproportionality in special education. Although recognizing the limitations of E&S data when not supplemented with information from other sources, the coeditors of this volume nonetheless argue that data on racial disproportion-

ality in special education and other facets of education are important and provide information that is essential both to efforts to improve the quality of education and to equitable access to it. Losen and Orfield (2002a, p. xxii) note: "In a society where race is so strongly related to individual, family and community conditions, it is extremely difficult to know what part of the inequalities are caused by discrimination within the school." Commenting further on studies in their volume that are based on E&S survey data and supplemented with information from other sources, the coeditors added: "These studies, however, do uncover correlations with race that cannot be explained by factors such as poverty or exposure to environmental hazards, alone" (Losen and Orfield, 2002a, p. xxii).

Like the 2002 National Research Council report, the coeditors of *Racial Inequity in Special Education* state that the papers in their volume also "suggest that special education issues faced by minority children often begin with shortcomings in the realm of general education well before teachers or parents seek an evaluation for special education eligibility. Therefore, policy solutions that fail to consider the connection with general education classrooms will unlikely bring about significant change" (Losen and Orfield, 2002a, p. xxiii).

---

[11]This report questioned both the validity and the significance of E&S findings of disparities, as well as those based on recent OSEP/OSERS surveys. The report argued that because neither survey provides important information about diagnostic criteria, students' needs, and other background information that could be used to determine the appropriateness of special education services to meet individual students' needs—regardless of their race—data from these surveys are not helpful in determining whether or not racial disproportionality in special education is actually a problem (National Research Council, 2002b).

In the absence of additional information from other surveys or from field research, the E&S survey data cannot by themselves prove whether racial differences in assignment to special education or other aspects of schooling that are addressed by the survey are the result of discrimination or constitute violations of students' civil rights. Indeed, the survey was never intended to be used as the sole source of information about discrimination.

3

Use of E&S Survey Data

As is true with all information resources, any assessment of the value of the E&S survey depends on the purposes for which the data from the survey are used. As discussed in Chapter 2, the E&S survey is the sole source of national school-level data about a number of issues related to equality of access to opportunity to learn. For some issues, it is the only available source of information; for many other issues, it is the only source of information that is current, disaggregated, and can be linked to specific schools and school districts. Yet officials of the Office for Civil Rights (OCR) of the U.S. Department of Education, which administers the survey, have stated that the E&S survey is underutilized, not only by researchers and the general public, but also within OCR itself.[1] OCR

officials say they would like to see the survey data used more extensively. In this chapter we examine how E&S survey data have been used to enforce civil rights laws, to promote public involvement in efforts to ensure that schools provide equal access to high-quality education, and as a resource for education researchers.

USE IN ENFORCEMENT

Internal OCR Use

With a budget of $71.2 million (in fiscal 2000), the OCR employs 750 full-time staff members located at the headquarters in Washington, DC, and in 12 regional offices throughout the

---

[1]There is little published information on how OCR uses E&S survey data. Information on OCR's use of the survey is derived primarily from

conversations and interviews with OCR officials and others who are knowledgeable about the survey and its uses.

United States. OCR is responsible for ensuring compliance with civil rights laws in higher education, in addition to elementary and secondary education (although the E&S survey and this report cover only the latter). More than two-thirds of the approximately 5,000 complaints it fields each year involve elementary and secondary schools (U.S. Department of Education, Office for Civil Rights, 2000a).

Since 1999, survey data have been available to OCR staff either over its private Internet-based network or on compact disks in a format that enables users to compile descriptive statistics and to combine data from different schools and districts. Although some lawyers in the regional offices often use E&S data in their work, most OCR staff seldom, if ever, use them. One reason for this lack of use is that OCR staff have not been provided with technical assistance or professional development on how to access and manipulate the data or on how the survey data could be used in enforcement activities (Peter McCabe, former director, E&S survey, personal communication, 2002).

Historically, OCR has been primarily interested in the collection of E&S survey data as records rather than in the analysis or dissemination of findings. The data, collected and certified by superintendents and principals or their designees, sometimes have been used by OCR officials as background information as they investigate complaints. Except for a brief period in the late 1970s and early 1980s when books containing descriptive statistics for all surveyed districts and schools were produced (e.g., U.S. Department of Health Education and Welfare, Office for Civil Rights, 1978), the only access to E&S data that OCR staff had was in the form of paper copies of all the compliance reports (i.e., the E&S survey forms) submitted by individual schools and districts. The reports were shipped to each of the 12 regional offices in cardboard boxes. In this format, the data could be used as background information while investigating complaints involving individual schools and districts, but for little else. OCR has not routinely analyzed or even compiled the data, although the data from each administration of the survey have been recorded and stored in a computer-readable format for every administration of the survey since 1968.

The most common use of the survey data by OCR is to identify schools and districts where discrimination may be occurring. As noted by OCR attorney Richard Foster (personal communication, October 2001), "If the data can get us closer to where discrimination is really occurring, then our enforcement work becomes more effective." Compliance reviews are selected partly on the basis of E&S survey data, as well as from information provided by parents, education groups, media, community organizations, and the public.

Compliance reviews are OCR-initiated investigations of state and local education agencies. The issues of concern and locations are selected on the basis of OCR's enforcement priorities and informed by data from the E&S survey. Compliance reviews are described as proactive efforts to address civil rights issues and generally involve practices that affect many more students than is typically the case when OCR responds to individual complaints (U.S. Department of Education, Office for Civil Rights, 2000a; Glennon, 2002, p. 213).

In 2000, 47 compliance reviews were initiated on the following topics[2]:

- ensuring that nondiscriminatory practices are followed in the placement of minority students in special education and in the provision of access to gifted and talented programs;
- ensuring that English-language learners are afforded access to

special language services in order to benefit from a school district's educational program;
- ensuring that students are not subject to a racially hostile environment;
- ensuring nondiscriminatory student disciplinary policies and practices; and
- ensuring equal opportunity for male and female students to participate in athletic programs.

That year, the largest number of compliance reviews was focused on the appropriate identification and placement of minority students in special education— collectively known as MINSPED (Glennon, 2002).[3] MINSPED compliance reviews are briefly discussed below to illustrate the role that E&S survey data have played in OCR enforcement activities.

Between 1993 and June of 2001, OCR initiated 167 MINSPED compliance reviews—most between 1995 and 2000. Schools and districts selected for MINSPED compliance reviews were identified primarily on the basis of statistics from the E&S survey—especially that of 1994. Nearly all of these

[2]The number of compliance reviews conducted varies substantially from year to year in accordance with changes in the Department of Education's administrative priorities. According to OCR staff, no compliance reviews were initiated in 2002.

[3]Theresa Glennon's account of the MINSPED initiative is one of very few recent published accounts of OCR enforcement activities.

compliance reviews resulted in voluntary agreements in which school districts agreed to put in place a number of measures to "reduce the misuse of special education for students of color" (Glennon, 2002, p. 199). Most of these agreements focused only on procedures for the appropriate referral, evaluation, and placement of minority students to special education programs. As Glennon (2002, p. 205) notes:

> This single-issue focus may be helpful in some instances, but it could also undermine OCR's ability to direct school districts toward more comprehensive reforms in both general and special education—reforms that may be necessary to resolve the overrepresentation problem, given the evidence that multiple factors contribute to this problem.

This point also has been made by Hehir (2002), Losen and Orfield (2002a), and the Committee on Minority Representation in Special Education (National Research Council, 2002b) in regard to OCR enforcement in the area of special education. Halpern (1995) has expressed similar views on the limitations of addressing pervasive racial disparities in education solely through civil rights enforcement efforts, as narrowly construed.

However, some agreements transcended a narrow focus on ensuring that the students' rights under the laws governing special education are strictly enforced. Some also addressed deficiencies in other school programs that may contribute to the observed racial disparities in special education. On occasion, the agreements that OCR negotiated with districts also addressed racial disparities in disciplinary practices, resource inequalities, achievement gaps on high-stakes tests, underrepresentation in gifted and talented programs, and overrepresentation in low-track courses of study (Glennon, 2002).

Glennon cites a 1999 report from OCR's Chicago regional office on its MINSPED compliance reviews as an example of how OCR's enforcement efforts can produce more comprehensive approaches to addressing the problem of inequality of opportunities to learn. The Chicago office reported that its enforcement efforts were more effective when they "included an emphasis on improving the ability of a school district's reading programs to serve its diverse student populations" (Glennon, 2002, p. 206).[4]

---

[4]As noted by the Committee on Preventing Reading Difficulties in Young Children (National Research Council, 1999b), more than 2 million children have been diagnosed with reading disabilities and are enrolled in special education for that reason. Students identified as reading disabled constitute approximately 80 percent of all children receiving special education services (see also National Research Council, 2002b).

Although she recognized the promise of some of the agreements that were negotiated as a result of MINSPED compliance reviews, Glennon (2002) noted that OCR does not publicize the agreements nor has the office evaluated their effectiveness. She also observed that none of the agreements included numerical requirements for change. However, even if the agreements included quantitative goals, OCR's data system is not prepared to monitor compliance because in most years the E&S survey is not administered to all schools, and no data are available for schools not included in the survey. Assuming that a school district is targeted for a compliance review in part on the basis of E&S survey data, there is no guarantee that that district will be included in the next administration of the survey.[5] Nevertheless, OCR claims that carefully targeted "compliance reviews nearly always result in recipients (of federal financial assistance) making policy or program changes that benefit large numbers of students— unlike complaints where remedies may benefit only the complaining party" (U.S. Department of Education Office for Civil Rights, 1999).

Increased use of E&S survey data for enforcement by the OCR is contingent on factors related to the administration of the survey, as well as OCR's enforcement policies and priorities. OCR would need to provide technical assistance to its staff in the 12 regional offices on how to access and manipulate the data. Also, professional development related to the potential uses of survey data in education civil rights enforcement would be needed.

The extent to which OCR chooses a proactive enforcement strategy by engaging in compliance reviews affects the probability that OCR staff will use E&S data, and the number of compliance reviews initiated since 2000 has decreased markedly.

E&S survey data could also be used to monitor the nonbinding partnership agreements, which have become the predominant OCR strategy to promote compliance with the civil rights laws (U.S. Department of Education Office for Civil Rights, 2000a). These agreements typically do not have numerical goals (Glennon, 2002), and E&S survey data have not been used to monitor outcomes (personal communication, Rebecca Fitch and Peter McCabe, Office for Civil Rights, 2002). Because OCR does not publicize the agreements it negotiates, there is no way of knowing what changes occur as a result of its enforcement activities. If OCR agree-

---

[5]Despite this limitation, the Office for Civil Rights planned to integrate data from the E&S survey into its Case Information System (CIS II) in early 2003 (Peter McCabe, personal communication, 2002).

ments were made public and included quantitative goals, the E&S survey could be an important resource not only for monitoring compliance, but also for measuring the outcomes of OCR enforcement efforts. For this to occur, the school districts with which OCR has negotiated partnership agreements would have to be routinely included among sampled districts in each administration of the survey.

Litigation

Exactly how often E&S survey data have been used by the public in litigation over students' civil rights is unknown, but civil rights advocates estimate that E&S survey data have informed dozens, and possibly hundreds, of lawsuits over the years. The lawsuits have addressed issues of equitable access to public education for minority children and youth, those whose first language is not English, and children with disabilities. An important example of the use of E&S data in litigation occurred in the 1974 Lau decision of the U.S. Supreme Court; the data had been used extensively by the U.S. Commission on Civil Rights in a series of reports on the education of Mexican Americans (see Sotomayor, 1974). The reports were cited by the Supreme Court as an important basis for the recognition of the constitutional rights of Hispanic students (*Lau v. Nichols*, 414 U.S. 563, 1974).

However, except for issues related to special education, litigation of education civil rights issues has become increasingly rare in recent years (Ryan, 2002; Paul Weckstein, director, Center for Education Law, personal communication, October 2002). For example, in the 1960s, the federal government provided funding to the Harvard Center for Law and Education to conduct analyses of education civil rights issues and to bring litigation to protect students' rights. Federal funding supporting education civil rights litigation did not continue into the 1970s, although foundation support for such organizations as the NAACP Legal Defense Fund and the Mexican-American Legal Defense and Education Fund, which still include litigation among their civil rights advocacy activities, has continued to varying degrees. In addition to the withdrawal of federal funding, education civil rights litigation was affected by court rulings that have made it more difficult to bring class action lawsuits.

Courts have been increasingly reluctant to accept plaintiffs' claims based on the disparate impact clause of Title VI (Ryan, 2002). Although there are many educational policies and practices that have a racially disparate impact, Title VI disparate impact claims have been raised primarily in challenges to testing, tracking, and funding decisions that have a racially disparate impact (Ryan,

2002). The first step of proving a racially disparate impact is usually not difficult. For example, plaintiffs making disparate impact claims related to high-stakes testing must first show that the use of a test has a disparate adverse impact on a protected group. Evidence of statistical disparities, such as those often found in analyses of E&S survey data, typically is sufficient. Once the plaintiffs meet this evidentiary burden, however, defendants have the opportunity to demonstrate that the practice contested (e.g., a test) is an educational necessity—that it serves "the legitimate educational goals of the institution" (see *GI Forum*, 87 F. Supp 2d at 679, citing *Cureton v. NCAA*, 37 F. Supp. 2d 687, 697 [ED Pa. 1999]). Despite rather specific and stringent professional standards for fair and appropriate testing (American Educational Research Association, 1999; National Research Council 1999b), courts generally defer to school leaders' claims about the educational necessity of testing and other challenged practices (Ryan, 2002). Thus, data documenting disparate impacts of various educational practices that can be derived from the E&S survey are relevant but generally have not been decisive.[6]

Finally, it should be noted that in 2001, the Supreme Court ruled in *Alexander v. Sandoval* that private litigants may no longer rely on Title VI's implementing regulations in cases presented in a formal court of law, although they may continue to file complaints based on Title VI with the OCR (Losen and Welner, 2002).

Except for a brief period during the late 1970s and early 1980s (U.S. Department of Health, Education and Welfare, 1978), the OCR has never published or broadly disseminated findings from the E&S survey. Data were made available on request to the public. Generally, requests for data have been for topics covered by the E&S survey for specific schools and school districts. There is no record of the number of requests for E&S data over the years, but it appears that E&S data were frequently requested and often used in school civil rights litigation during the 1970s and into the 1980s (Halpern, 1995).

By the 1990s, however, the E&S survey had become close to invisible to most outside of the Department of Education, except for a small number of researchers and education advocates.

---

[6]For a more complete discussion of criteria by which disparate impact claims concerning high-

stakes testing have been made, see National Research Council (1999c). For further discussion of litigation regarding racial disparities in special education, see Losen and Welner (2002).

In 1996, OCR received about 15 requests per month for E&S survey data. The survey was so little used that in 1996, Norma Cantu, the Department of Education's assistant secretary for civil rights, had tentatively decided to terminate it. However, after extensive discussions that included education advocacy organizations, she reversed her decision, and OCR made a commitment to update the content and improve the administration of the survey "to make it more useful" (Peter McCabe, Office for Civil Rights, personal communication, 2002). Among the administrative objectives was to reduce the amount of time required to make the data available from 18 months to 6 months and to make the data more readily available to the public.

Another strategy to make the data more useful was to administer the survey in all school districts in the United States for only the second time in the history of the survey. This was done in 2000 to coincide with the administration of the census. OCR officials hoped that researchers would link E&S school data with neighborhood population data corresponding to school attendance boundaries, as well as to other datasets. It was hoped that researchers would then conduct analyses that would provide incisive and comprehensive overviews of the current state of education civil rights issues in the United States and how various school and nonschool factors interact in affecting equality of access to educational opportunities.

By the spring of 2002, the number of requests to OCR for E&S survey data had more than doubled, to approximately 35 per month. More than half of these requests were coming from the press. Education advocacy groups, such as the National Coalition of Advocates for Students and its local affiliates, increasingly were informing the press about the existence of E&S survey data that were pertinent to local school problems that were receiving attention.

Although demand for E&S data was increasing, OCR was still far from achieving the kind of visibility and access to the data that it desired. Informing the public about students' rights is viewed by OCR as an important part of its mission and an essential tool for ensuring equal access to educational opportunities (Richard Foster, Office for Civil Rights, personal communication, October 2001). To this end, OCR maintains a very detailed website (www.ed.gov/offices/ocr) describing the mission of OCR, the civil rights laws it enforces, and examples of how enforcement of those laws affects students' opportunities to learn.

In the summer of 2002, OCR made E&S survey data, including data for individual schools and districts, available to the public on its website. In August 2000, during its third month of operation, the civil rights data page on the OCR website was accessed 1,858 times (Bernadette Adams-Yates, Office for Civil Rights, personal communication, 2002). No information has as yet become available on how these data are being used. (Although OCR's efforts have made the data much more accessible, the software involved could be improved, a point we discuss in Chapter 4.)

PUBLIC USE

Another important use of E&S survey data is to promote informed democratic involvement in public education. To this end, education advocacy organizations such as the National Coalition of Advocates for Students (NCAS) use E&S survey data to inform and mobilize citizens' involvement in their schools. Long before the E&S survey data became available over the Internet, organizations such as NCAS obtained the data from OCR to produce publications that inform parents and families about the level of access available to their children in local public schools (see Carmona, Wheelock, and First, 1998; National Coalition of Advocates for Students, 1998).

In fact, NCAS, its affiliates, and other child advocacy organizations have been the primary consumers of OCR survey data over the years. Their publications include data-based education advocacy reports that contain qualitative and quantitative information about student achievement and school exclusion and how they relate to each other. These reports also compare the school experiences of minority children with those of their white peers.

Advocacy groups have also provided the data to the press, who often write about the implications—including unintended negative consequences—of local school reforms. For example, NCAS has given priority to teaching local education writers how to access E&S data and use survey findings to inform substantive stories about local public schools. In September of 2002, NCAS mailed press kits to more than 600 members of the Education Writers Association that included Internet addresses for 2000 E&S district and school-level survey data. The writers were encouraged to contact NCAS and its member organizations to discuss the implications of these data. Similar efforts are made by other advocacy organizations, such as the Applied Research Center (see Gordon, 1998; Gordon, Della Piana, and Keleher, 2000; Johnson, Boyden and Pitz, 2002;).

An example of how widespread reporting of disaggregated data from the E&S survey has influenced school policies comes from Florida in the 1980s. When Florida media, using E&S survey data, reported that many Florida school districts routinely administered corporal punishment to 20 percent of their black students each year, a number of districts changed their disciplinary practices, and the frequency with which corporal punishment was used dramatically decreased (Joan First, director, NCAS, personal communication, 2002).

RESEARCH USE

Until recently, E&S survey data rarely have been used for academic research. A few researchers have used the data to analyze issues related to school desegregation (see, e.g., Farley, 1975, 1976, 1978; Farley, Richard, and Wurdock, 1980; Welch, 1987; Farley and Taeuber, 1974; Orfield, 1977, 1978, 1986, 1996; Orfield and Yun, 1999). The data also have been used to discuss racial disparities in special and gifted education (see, e.g., Harry and Anderson, 1995; Ford, 1998; MacMillan and Reschly, 1998; Oswald et al., 1999; National Research Council, 2002b). Recently, a number of publications associated with the Civil Rights Project at Harvard University also have used E&S survey data (see, e.g., Losen and Orfield, 2002a, 2002b; The Advancement Project and The Civil Rights Project, 2000; Orfield, 2001b).

Because the OCR until recently has not publicized the survey nor disseminated even basic findings, many researchers have been unaware that the survey data exist. Some researchers who have been aware of the survey have had difficulty gaining access to the data in a format that allows for the use of a full range of analytical strategies. When access has been obtained, technical documentation has been sparse (see Chapter 4).

4

Strengthening the E&S Survey Data

The primary purpose of the E&S survey has been to provide the Office for Civil Rights (OCR) with data to use when targeting compliance review sites or when investigating complaints. Because OCR attempts to settle complaints through informal procedures and rarely issues findings as part of a response to a complaint, it is unclear whether and to what extent OCR or complainants use the data.

The lack of strong information about use raises the issue of the basic purposes of the E&S survey and what can be done to strengthen it, both technically and to address policy concerns. Although the committee recognizes that the original intents of the survey continue to be important, many educational and societal changes have occurred since its inception. The state of the art in data collection, analysis, and coordination with other data systems has also advanced greatly since the inception of the survey. These factors, plus the passage of the No Child Left Behind Act of 2001, which includes substantial accountability provisions for states and local school districts, argue for expanded availability and use of the E&S survey data.

The No Child Left Behind Act focuses attention on the outcomes of educational practice by requiring testing in reading, mathematics, and science and related accountability for results and by requiring that the results be disaggregated by several categories, including gender, race, ethnicity, English proficiency, disability status, and economic disadvantage. The E&S survey could help provide a fuller picture of the educational process with information on educational inputs that describe the access of students to quality opportunities to learn. Halpern (1995) suggests that OCR has strongly focused on frequency counts of racial categories and that a reliance on this approach will have limited value unless it is supplemented by a focus on the quality of the educational opportunities that students experience.

This chapter highlights ways in which the E&S survey could be strengthened and therefore be made more useful to OCR and others concerned with ensuring access to learning opportunities. The committee offers several ways to strengthen the survey in three broad categories: methodology and technical issues, content, and use.

METHODOLOGY AND TECHNICAL ISSUES

Field Testing and Respondent Validity Studies

The E&S survey instruments are somewhat complex and require respondents to collect a substantial amount of detailed information—such as on enrollments and dropout rates, children with disabilities, racial and ethnic categories, disciplinary events, testing, student assignment, athletics, and teacher certification. The capability of districts and schools to collect the required information efficiently and accurately varies greatly. Some districts have computerized student identification and data management systems in which most of the needed information is routinely collected and analyzed, while other districts and schools may use a paper system to collect some or all of the data. Still others collect the data that the state requires once a year for fall enrollment procedures and do not collect any additional information. In addition, as noted above, several of the questions may not be easily understood by respondents or ask for information that could be understood or interpreted differently by different jurisdictions.

In addition to issues related to the various data collection systems is the question of which employees at the school or district level actually complete the survey. The committee heard from members who have responsibility for data collection in their jurisdictions that there are substantial differences in roles and responsibilities among respondents. In some cases, a data manager may complete the forms; in other cases, an administrator (e.g., school principal or assistant principal) is charged with the responsibility. In still other instances, a clerk from the central office of the district or school may be given the task. A strength of the survey is that it requires a signed certification that the data are accurate. While this requirement enhances the probability of accurate data, it does not necessarily ensure that the data were collected and reported in a consistent and reliable manner.

These variations in the way the E&S survey is administered can be expected to have an effect on the overall reliability of the information collected. The degree of this effect is not known, but minimizing the unreliability of the information is critical. The committee suggests two courses of action. First, there should be an extensive field-testing component as changes are made in the survey, as is done for state and local student testing programs that regularly conduct item tryouts, and field testing of new and revised assessment instruments to be sure that students can understand the test questions and respond appropriately. Second, OCR should consider more extensive field testing of its new or revised forms to help ensure that respondents understand the nature of the questions and how to complete the forms. Similarly, OCR should also conduct validity studies to determine whether the information being collected is, in fact, valid. This kind of study would require that schools be sampled and the information they supply be compared with documentation that exists. Discrepancy rates could then be calculated and the instances in which no backup documentation exists could be tabulated. This type of validity study would give OCR an idea of how accurate the results from the survey really are. If it is not possible to conduct a full-scale validity

study, OCR should consider implementing a recommendation from Croninger and Douglas (2002) to conduct a series of small-scale focus interviews to determine how school and district administrators complete certain key tables in the survey.

Tracking Trend Data

One common purpose for large data collection systems is to provide trend information over time. It is certainly logical to expect that a survey that has existed since 1968 would have some capability to provide long-term trends. At present, this capability seems quite limited. The data are not accessible in a common electronic format and therefore cannot be easily retrieved or manipulated. Both the computer tapes and printouts of descriptive information from surveys administered from 1968 to 1992 were not stored in a central location and effectively were lost for several years, then relocated in 1998. Although the data have been transferred from the tapes to computer disks, the data for various survey administrations are in incompatible formats, as they were compiled by computer software and hardware that are now obsolete. For this reason, OCR currently cannot access and analyze most E&S data longitudinally (Peter McCabe, former director, E&S survey, personal communication, 2002).

For historical purposes, OCR should undertake efforts to compile past data for comparable survey questions since its 1968 inception. OCR also should consider the benefit of building a data system that will strengthen the capacity to allow for easily generated, accessible trend data for reporting categories in the future.

Data Projection Methodology

A technical report on the E&S survey done for OCR (U.S. Department of Education, Office for Civil Rights, 2000b) addressed the issue of state and national projections from the reported E&S survey data. Depending on the years, the survey samples varied in their statistical validity. For most years, the samples have been of sufficient quality to allow for projections to the state and national levels. However, the surveys collected in 1969, 1971 and 1973, 1982, and 1990 did not yield statistically valid samples, so defensible projections could not be constructed.[1] Also, other problems with the survey administered in 1996–1997 made these data unreliable (Glennon, 2002, p. 213; Peter McCabe, personal communication, 2002).

OCR should routinely utilize appropriate sampling methodology techniques to ensure that projections to state and national levels can be produced. In recent years, the projections, when they have been calculated, have provided a "best estimate," but the methods that are currently used do not allow calculation of confidence intervals to clearly identify the degree of error associated with the projections. In addition to developing the methodology to routinely provide projections, OCR should produce and report confidence levels.

Adjusting Protocols

The committee discussed a number of issues related to the changing demographic nature of the country and the different initiatives that states and schools have implemented to address the situation. As an example, the committee finds that the collection of classroom-level data (Item 13) could be a major strength of the survey, but that problems with the protocol for data collection and ambiguity in the wording of the question pose problems. Here we address the protocol issue; the problem of ambiguous wording is addressed separately, below.

The survey protocol has historically placed limits on which schools complete the classroom portions of the survey. OCR collects classroom-level data at the entry and exit grades (e.g., grade 1 and

---

[1]According to the technical report (U.S. Department of Education, Office for Civil Rights, 2000b), the survey for 1990 produced a sample that could be projected to the national level but was not valid for state-level projections.

grade 6) for elementary school pupils in schools in which minority enrollment in the school is greater than 20 percent and less than 80 percent. Classroom-level data collected for these schools include the grade level, ability grouping, and number of students by race, ethnicity, and English proficiency.

The data collection procedure used by OCR eliminates elementary schools that have fewer than 20 percent or more than 80 percent of their students from racial and ethnic minority groups, as well as all middle schools, junior high schools, and high schools. The committee notes that this protocol might have made sense when the demographic composition of the nation and its schools was less complex, but it is inadequate for the multiethnic demographics of schools in the 21st century. OCR should consider changing the data collection protocols so that all elementary schools selected in the sample would supply classroom data.

Data Editing

When administering any large-scale survey, such as the E&S survey, data editing is important. Because the administration and operation of the survey have not been well funded, data editing has frequently suffered. Glennon (2002) indicated that the data from the 1996 survey, actually administered in 1997, turned out to be unusable.

Officials responsible for the survey have indicated that there were insufficient funds to do proper editing of the data that were received and that, in times of budget shortfalls, data editing is frequently a casualty.

Croninger and Douglas (2002) conducted analyses to look at the prevalence of high-stakes testing. They found that the data they used from Tables 12A and 12B of the E&S survey for 2000 had not been adequately edited or verified by OCR. They had to make certain logical assumptions about treating the data to do their analyses; these inferences could have been avoided had proper editing been done and documentation provided. Even when data editing has been done, the contractor has in the past carried out the editing with little or no oversight from OCR. In the future, OCR should place a priority on editing the data that are collected in order to have the best datasets available for use by OCR, advocates, parents, and researchers.

CONTENT

Changes in Existing Survey Items

In fulfilling its charge, the committee examined the survey items to determine whether improvements to specific items would enhance the validity and reliability of the survey. The committee identified seven topics as prime candidates for revision: children with disabilities, testing, high school completion, student assignment, advanced placement classes, interscholastic athletics, and teacher qualifications.

Children with Disabilities

Item 10.1 (Table 10.1) collects information on children with mild, moderate, and severe retardation by race and ethnicity. Item 10.2 (Table 10.2) collects information on children identified as emotionally disturbed and with specific learning disabilities, by race and ethnicity. Both items also ask for information about the educational placement of these students (i.e., the percentage of time spent outside of a regular classroom), but no data on the race and ethnicity of students in different educational placements are requested. These items would be strengthened by collecting race and ethnicity data for educational placement.

How students with disabilities are served makes a huge difference in their future prospects (see Vaughn et al., 2000; Swanson, 1999; National Research Council, 2002b, pp. 324–328), and the quality of services available for children with disabilities varies widely. In many cases, students are well served. But, not infrequently, students are identified as having disabilities and sent to groups or classrooms that actually reduce their learning opportunities (see Garcia Fierros and Conroy, 2002). Misassignment can be the consequence of teachers' inability to serve difficult-to-teach students (see Harry et al., 2002). There is abundant evidence that black, American Indian, and Hispanic students are disproportionately identified as having certain types of disabilities (Finn, 1982; National Research Council, 2002b; Losen and Orfield, 2002b).

Testing

High-stakes tests are those whose passage is required before a student is allowed to advance to the next grade or graduate from high school. Retention in grade often results in increasing the likelihood that students affected will drop out of school or learn less than if the need to accelerate their learning had been addressed in other ways (National Research Council, 1999c, pp. 128–132). Item 12 asks about the use of high-stakes tests for grade promotion and high school graduation.

Tables 12A and 12B ask whether testing information is used as a "sole" or "significant" criterion for grade promotion and for high school graduation, respectively. The meaning of "sole criterion" and "significant criterion" is ambiguous, and item 12 would be improved by clarifying the meaning of these terms. In addition, the item does not ask for the number of students who fail the test the first time. As a result, the current information contains repeat test takers who have failed one or more times; this confounds the results, and the meaning of the data becomes unclear. The importance of improving this question is heightened by the provisions of the No Child Left Behind Act, which mandates increased testing during the elementary and middle grades.

Student Assignment

As noted in the discussion of protocol, above, Item 13 elicits information on the classroom assignments of students from different racial and ethnic groups and for students with limited English proficiency. The item also asks whether any students are "ability grouped for instruction in mathematics or English-Reading-Language Arts" in that classroom. Information on classroom assignment is requested for the lowest and highest elementary grades only.

The civil rights concerns emanate from evidence that many students who are "tracked" on a continuing basis into separate classrooms or groups within classrooms because of their below-grade-level performance continue to lose academic ground in these settings, that minority students are disproportionately "tracked" into low-ability classes, and that such practices may produce "within-school segregation"

(see Oakes, 1990; Mickelson, 2001). Experts agree, however, that some approaches to ability grouping (or, more accurately, performance grouping) can serve important educational purposes providing they enlarge, rather than restrict, opportunities to learn (see Slavin et al., 1994). So knowing that students are grouped by ability, by itself, is not sufficient to identify discriminatory or ineffective practices.

Also, the question does not ask how many students of each racial or ethnic group are in the upper or lower ability groupings. The definitions and instructions for ability grouping are complicated and could be confusing to respondents at the school level. Specifically, student assignment data are organized according to teacher identification codes. This can be ambiguous, particularly in the upper elementary grades in which students are more likely to be taught by more than one teacher. Also, as noted in the protocol section above, the fact that only schools in which minority students constitute more than 20 percent but less than 80 percent of total enrollment are required to complete Item 13 is problematic.

Information about classroom assignment and ability grouping is extremely important, but the current wording of this item severely limits its value. This item could be greatly improved with minor modifications: providing a clearer definition of ability grouping, the addition of information about the subject matter being taught (e.g., Is it a language class for English-language learners?), as well as information about the racial and ethnic composition of students in the highest and lowest groups. In addition, all sampled schools should be required to provide information on classroom assignment.

Advanced Placement Classes

When students are motivated to learn, the opportunity to engage rigorous curricula often leads to higher achievement (Adelman, 1999). The absence of such learning opportunities restricts what and how much students learn and gives an advantage to those students who do have access to more demanding courses and programs. Item 14 asks about advanced placement (AP) classes offered by the school. However, the survey does not provide a denominator that is more specific than the overall racial and ethnic composition of the school. For example, information is not cohort or grade specific. This makes estimation of the percentages of students in each race and ethnic group who are in AP classes less precise than it might otherwise be. Also, the question is limited to AP classes and does not collect information on other advanced courses of study like the international baccalaureate program and honors

programs. The item should be expanded to include other advanced study programs and data specifically on groups by grade or cohort.

High School Completion

Graduation from high school is, of course, a critical step toward college or well-paying jobs. Item 15 asks about high school completion (diploma and certificate of attendance or completion) offered by the school. The race and ethnicity data do not provide specific denominators (e.g., the number of students from each race and ethnic group entering high school), so it is not possible to determine the percentages of students who graduate from each race and ethnic group. Also, there seems to be considerable ambiguity in the meaning of the types of completion certificates. For example, some states have begun to offer a Certificate of Initial Mastery and Certificate of Advanced Mastery to their students, a trend that may accelerate as a result of the No Child Left Behind Act of 2001. Gathering data specifically on groups by grade or cohort as well as clarifying the definitions for the type of completion certificates would strengthen the item.

Interscholastic Athletics

Item 16 asks about the number of different sports and teams offered at the school and whether the sports that are offered include males only, females only, or both male and female participants. The item elicits useful information on gender equity, but adding information on the race and ethnicity of students would be useful. Although many school districts under long-standing school desegregation court orders have been required to demonstrate equitable access to extracurricular activities for black students, data have not been routinely collected to allow comprehensive monitoring. The E&S survey has collected information on student participation in interscholastic athletics since 1994 to monitor Title IX issues, but information on the race and ethnicity of student participants has not been collected for Title VI purposes. The collection of this information either in this question or as an additional item would provide useful data to OCR and those concerned with equity in extracurricular learning opportunities.

Teacher Qualifications

The school-based learning opportunity that accounts for the greatest variation in student achievement is quality teaching (see Sanders and Horn, 1995; Sanders and Rivers, 1996; Hedges, Layne, and Greenwald, 1994; Darling Hammond, 1997a; Ferguson, 2000). To measure this, Item 17 asks how many full-time teachers employed by the

school meet requirements for a standard certificate. With the field of teacher certification rapidly changing in states and the provisions of the No Child Left Behind Act that every child have a fully qualified teacher, the committee questions whether the concept of a "standard certificate" has become ambiguous. The language of the question should be clarified or possibly changed to include whether teachers are teaching in the content field or specialty for which they were trained to teach.

A related issue with respect to teachers is their years of teaching experience. Teacher inexperience is negatively related to teacher effectiveness, at least in the first 3–4 years of teaching. Since minority students, those with disabilities, and those with limited English proficiency are often more likely to be taught by novice teachers, teacher experience should be a subject of the E&S survey.

Items That Might Be Eliminated

The committee discussed whether some items could be deleted to help streamline the survey and to rid it of questions that are not actionable or cannot be cast into language that would elicit clear and useful responses. Items 6 and 6a on the District Summary Report (ED 101) cover how many students were identified as pregnant during the previous academic year and then how many currently are not in school. The committee concludes that these questions are not particularly helpful in generating actionable data for OCR enforcement, nor do they connect directly to civil rights concerns.

USE

Improving Access to Survey Data and Survey Findings

The OCR E&S survey has been useful to a wide variety of users, including state and federal agencies, education advocates, civil rights attorneys, and academic researchers. Although the data have been used for important purposes, they could be more widely and productively used.

Historically, the data files have been difficult to access and utilize in complex examinations. Recently, however, OCR has made some significant strides in making information more accessible to the public and to education advocates. OCR has placed some of the information on its website so that users can access the data and query them in simple ways. This action should make the information much more useful to the general public, and OCR should be commended for taking this step.

The fledgling effort should be evaluated to determine whether the visitors to the website find that the information and the formats provided are useful. It would also be helpful to find out whether users are content with the level of analysis currently available or whether other, more detailed analyses are desired. Also, given the sporadic nature of OCR data editing of the survey, there should be an examination of possible data errors and whether, if errors are found, it is a substantial enough problem to discourage the public from using the data.

There are a number of actions that OCR could consider that would improve access to the survey data. Several of these possible solutions would require a greater allocation of resources within OCR. Some actions, however, may not require additional monetary resources but rather would demand more coordination and cooperation between offices within the Department of Education.

Formatting the Data to Make It Easier to Use

Some of the academic researchers commissioned by the committee (e.g., Ready and Lee, 2002; Croninger and Douglas 2002; see Appendix A) found the files to be problematic to use, prompting several attempts to have the datasets provided in different formats.

Eventually some of the researchers figured out ways to use the files or were provided with "flat files"[2] for certain years, which included much of the raw data that could be more easily transformed into other statistical formats for analysis. The process was not a smooth one and required the researchers to spend much time struggling to make the system work.

For research purposes, E&S data should be provided either as flat files with detailed codebooks or as well-labeled statistical package files (e.g., SPSS, SAS, or STATA). If the data are made available as any one of these types of files, researchers can then use transfer software (STATTRASFER or DBMScopy) to translate the files to any other package. Good documentation of such procedures is essential.

The National Center for Education Statistics (NCES) is the primary statistical agency for the collection, analysis, and publication of education data. Its goal is to collect, analyze, and disseminate statistics and other information related to education in the United States and in other nations. NCES has a large portfolio of data collection projects, including surveys in early childhood, elementary and secondary education, international indicators, postsecondary

issues, and the National Assessment of Educational Progress. NCES is a major source of educational data and information for the public. NCES has the professional staff and the experience needed to anticipate problems and issues that may arise as users of the E&S survey data attempt to secure and analyze the data. OCR should discuss with NCES the possibility of including E&S survey data in the datasets it makes available to the public.

OCR should also investigate whether selected findings could be published in other NCES or Department of Education documents that routinely get wide dissemination to states, school districts, and the public. One example of such a document is the department's annual *Condition of Education* report, which presents key findings from a wide range of data collection vehicles.

Training and Support of the E&S Survey Users

It is not unusual for users of large datasets, like the E&S survey, to need training and support to make maximum use of the information. For some users, it may be enough to access the less complex data displays that currently exist on the website. For more experienced users of the basic data, as well as researchers who want to do complex secondary analysis of the datasets and link them to other, non-OCR datafiles,

---

[2]A flat file is a text file that is not tied to any particular computer program.

additional support may be warranted to facilitate the work. Obstacles may include such topics as how to treat missing data, how to avoid potential misinterpretations of the data resulting from nonobvious definitions of the variables, and inaccuracies. Online tutorials and hard-copy data manuals are possible solutions to making the data more useful to a wider range of analysts. NCES may be able to provide some assistance to OCR since that agency has a solid history of providing web access and data analysis tools for its products. NCES has also conducted data analysis workshops for state education agency personnel, and university and private-sector secondary researchers, an activity OCR should consider.

Connections to Other Data

The E&S survey is but one of many surveys conducted by the U.S. Department of Education. The E&S survey serves a unique set of purposes, but it also contains questions that other surveys ask in similar, if not identical, ways, particularly for information on special education students because the department's Office of Special Education and Rehabilitative Services collects extensive data on services. The NCES collects education data for a wide variety of purposes, some of which overlap with the content of the E&S survey. OCR should work with these agencies to identify overlap and to see if redundancy exists. Moreover, as previously noted, Halpern (1995) suggests that OCR does not adequately measure curricular and programmatic changes that may be connected to discriminatory practices; rather it emphasizes collecting frequency counts. On its own, the E&S survey clearly cannot collect in-depth information on the quality of curricula and opportunities to learn to produce a full picture, but data from it could be integrated with other data collection efforts to achieve that goal (see Appendix D for a further discussion of this use).

Links to the PBDMI

As mentioned earlier, there is a new major effort under way in the Department of Education to consolidate the collection and maintenance of administrative data used for program management and policy. The initiative, known as the Performance-Based Data Management Initiative (PBDMI), began with top management support from major offices in the Department of Education, including OCR and the Offices of Elementary and Secondary Education, English Language Acquisition, Vocational and Adult Education, and Special Education and Rehabilitative Services. The intention is to lay out the information needed by each program against the statutory, regulatory, and other required information to ensure that only critical information is identified. The PBDMI will be launching many activities, including a demonstration project designed to provide a capability that links the department's various sources of state demographic, academic, and funding information together to support educational performance and achievement analysis. The plan is to transform the current data gathering process, which has numerous and sometimes duplicative collections, into a series of state-federal data exchanges with a central data repository. The PBDMI is expected to achieve partnerships with state systems beginning in 2003 and, given sufficient funding, full implementation by 2005.

The PBDMI effort is consistent with previous recommendations to OCR. For example, the National Research Council (2002b) recommended that the Department of Education conduct a single, well-designed data collection effort to monitor both the number of students receiving services through the Individuals with Disabilities Education Act and the characteristics of those children of concern to civil rights enforcement efforts. The PBDMI has the opportunity to be cognizant of the value of collecting programmatic data from schools on access to high-quality educational services and resources and to ensure that the public has access to that information. How the E&S survey goes through revision and continuation, while coordinating and cooperating with the PBDMI, will be critical to its future utility.

Implementation of the PBDMI is likely to affect the E&S survey, although precisely how is still unclear, since it is still under development. OCR should be sure to participate fully in the PBDMI discussions and ensure that the goals of the E&S survey data collection are represented in the implementation of PBDMI. PBDMI, when fully operational, may offer a unique opportunity to portray E&S survey data in a way that enriches both the OCR data and key educational information collected by the department.

Analysis and Dissemination

It would be very useful to those concerned with the provision of educational opportunities to minority students, to students with disabilities and those with limited English proficiency, and to advocates of gender equity to have easy access to simple tabulations of the data from the E&S survey. The publication of such work may encourage researchers to conduct more extensive analysis of the data. The publication of selected data should be in both electronic and print forms to allow for maximum access and dissemination.

Components of the survey could be disseminated in various ways. In addition to the suggestions above that selected findings be published in the department's annual *Condition of Education* report and that survey data might be part of the PBDMI process, OCR should consider requesting funds to continue and expand its data reporting efforts so that it could at least conduct simple analyses of the survey data during the year they become available. OCR should also consider publishing the findings in ways that would allow people to examine trends over time. Also, with the emphasis now being placed on the accountability provisions of the No Child Left Behind Act, OCR should consider posting on its website a full report of tabulations of the data showing how opportunities to learn are allocated to students of different backgrounds. This would allow analysts to balance the testing outcome data from the act with solid information about access to learning opportunities and resources described by the E&S survey.

5

Improving the Survey and Its Use

The work of the Office for Civil Rights (OCR) in addressing violations of students' civil rights and combating discrimination continues to be important, and the E&S survey is a useful, albeit underused, resource in OCR's efforts to enforce civil rights laws. The information produced by the E&S survey facilitates the identification of disparities in learning opportunities associated with race, ethnicity, language, gender, and disability status. Importantly, the survey is the only available data source that can be used to identify problems not only at the national and state levels, but also in school districts, and in individual schools. The committee finds that the E&S survey continues to play an important role in protecting the rights of minority students, students with disabilities, students with limited English proficiency, and women. The E&S survey is the only nationwide source of much of the information it collects. The survey's capacity to identify interdistrict, intradistrict, and even intraschool disparities in learning opportunities makes it a unique resource that provides some, though not all, of the information needed for spotting potentially actionable discrimination and violations of civil rights.

The survey currently is not being extensively used by OCR in enforcement, except as one of a number of sources of information used to identify school districts for compliance reviews. The number of compliance reviews initiated by OCR varies considerably from year to year; in 2002, no new compliance reviews were initiated. The survey's most important use in recent years has been to provide parents and others with information on disparities in access to high-quality learning opportunities.

The committee also finds that the

survey and the manner in which it is administered can be improved to: (1) more effectively assess whether students have access to critical learning opportunities; (2) make the data more accessible to interested citizens, as well as to educators and policy makers; and (3) enhance its usefulness as a resource for research that could lead to school improvement.

Despite the many societal and educational changes that have occurred during the past 35 years, major disparities in opportunities to learn and in education outcomes persist, especially those associated with race and ethnicity—the original focus of the E&S survey. The timely collection and analysis of classroom-, school- and district-level data that can help identify educational policies and practices that may have inequitable, if not discriminatory, effects on students is no less important today than it was when the E&S survey was first administered.

LEARNING OPPORTUNITIES— THE KEY

In one way or another, virtually all civil rights related to education involve the assurance that students will have the necessary opportunities to learn. The definition of "necessary" is contested, but the idea of equal opportuni-ties to learn for persons of different backgrounds is the basis of contemporary understandings of rights. Legislation and court rulings addressing the needs of persons who do not speak English or who have disabilities have extended the protection of the law and gone beyond the goal of equality to guarantee the provision of needed services. School finance cases also have sought to broaden the definition of individual rights related to education beyond equality of spending per student to take into account the fact that some students, including but not limited to students who do not speak English and students with disabilities, need more services than others if they are to succeed in school (see Rebell, 2002).

Regardless of the race, ethnicity, socioeconomic status, or disability status of the students whose educational rights are at issue, the focus of concern is learning opportunities. If students with certain characteristics have differential access to learning opportunities in particular situations, it may signal discrimination. However, despite the centrality of learning opportunities to the protection of civil rights, there is no accepted or even widely discussed model that identifies the full range of opportunities to learn, their relative importance to student learning, and their interrelationships. Research that links data on patterns of access to learning opportunities and resources, such as those from the E&S survey, with other datasets that are used to examine the effects of various policies, practices, and resources, potentially could produce a clearer understanding of what kinds of disparities make a difference for students' learning outcomes (see Appendixes A and D). Not only would the development of such a model be important for the protection of civil rights, but it would also shape education strategies generally.

Children's learning is influenced by many experiences. Those that are available in school typically account for less than one-third of students' academic achievement, as measured by most common tests (e.g., see Halpern-Felsher et al., 1997; National Research Council, 1999a). Yet the focus of the E&S survey, and thus the focus of this committee's interests, are the learning opportunities and resources provided in schools or by schools. Trying to identify the full range of influences on student learning that might reasonably be the concern of those who would protect civil rights is daunting. Nonetheless, the E&S survey focuses on learning opportunities that are of critical importance. Continuing refinement of the survey and judicious additions to its scope would make it an even more substantial resource for enhancing the

education of students who have been the victims of discrimination or of inadequate access to learning opportunities and resources.

PROMISING NEW USES OF E&S SURVEY DATA

Integrating E&S Survey Data with Other Information Resources

As discussed in Chapter 4, planning is under way to integrate the E&S survey into a consolidated data system that would coordinate the collection and management of data related to the administration of various programs administered by the U.S. Department of Education. The effort to develop this system, known as the Performance-Based Data Management Initiative (PBDMI), is being conducted in partnership with state departments of education. Besides integrating the administrative data systems of the department, other goals of PBDMI include facilitating the integration of state and local education data with the department's administrative database to make the data more usable and accessible for all.

Integrating the E&S survey with the PBDMI would help to ensure that the definitions of items that appear on the

E&S survey are consistent with those on other department surveys. Departmental literature on PBDMI also promises that it will reduce the data collection burden on schools by eliminating redundancy among the various surveys currently administered. It also promises that PBDMI will facilitate efforts to ensure accountability.

When the PBDMI data system is implemented, it will be important to ensure that the attributes that make the E&S survey an essential resource for monitoring equality of access to learning opportunities and resources is retained. Written descriptions of PBDMI from the Department of Education do not currently mention the collection and integration of data on access to learning opportunities and resources as a feature of the new data system. It is essential that information on access to opportunities to learn be an integral part of the PBDMI system. The new data system must continue to provide information pertinent to students' civil rights under Title VI, Title IX, and Section 504. These data must continue to be disaggregated by race and ethnicity, traceable to specific schools and school districts, and accessible not only to departmental officials, but also to local educators and to the public. If some or all of the E&S survey were incorporated into the PBDMI, it

could become less burdensome and therefore more cost-effective to routinely collect E&S data from all schools.

The Survey as a Resource for School Improvement Efforts

Large disparities in education outcomes among students of different backgrounds persist, and OCR continues to find that violations of students' civil rights are not uncommon. Whether or not the disparities identified by the E&S survey are caused by violations of students' legal rights, the disparities evidence the failure of schools to provide equal access to high-quality education for students from all backgrounds. This information is relevant not just to civil rights enforcement, but also to broader based efforts to achieve equity and excellence. Providing educators, parents, and the general public with easy access information that suggests that their own schools may be failing to ensure equal access to learning opportunities can lead to greater public involvement in school improvement efforts.

As society and public school students become more diverse, issues of equity will become important in places that previously have not had diverse student populations. The successful implementation of school reforms in a post-desegregation era will require that policies and practices be closely monitored so that they do not disadvantage minority students. The E&S survey provides a resource that can help to ensure that no child will be left behind. Without such a resource, education policy makers and practitioners may be unable to determine if reforms differently affect educational outcomes for racial and ethnic minority, language minority, or disabled students or even whether outcomes may vary by gender.

Until the summer of 2002, OCR had not published or otherwise made available to the public E&S survey data for more than 20 years—except by request. The recent placement of E&S data on the Internet is a very important innovation whose implications for public use of the data are not yet known. The software used to examine the data for exploratory purposes requires no technical training to use, and it allows individuals to make very specific queries about the various items addressed by the survey. For example, one could examine the number and percentage of sixth graders at a specific school who passed or failed a district- or state-administered test required for grade promotion, with the information disaggregated by race, ethnicity, and sex. Data for different schools can be compared, as can data for different states. Although some of the data from the survey are suppressed to protect stu-

dents' privacy, the data available over the Internet provide extensive information concerning disparities in access to learning opportunities.[1] Making these data easily available to the public is consistent with current federal and state policies emphasizing the public reporting of school-level data as a means of promoting accountability.

Educational Equity Self-Assessments

One way to do this could be in the form of an Educational Equity Self-Assessment (EESA). An EESA is similar in concept to the Racial Justice Report Card developed by the Applied Research Center of Oakland, California (Gordon, Della Piana, and Keleher, 2002). Using E&S survey data, possibly in combination with data from other sources, OCR could develop an EESA that educators, parent-teacher organizations, and others could use to evaluate individual schools. The EESA would be a computer-based, interactive program with access from the Internet or delivered on CD-ROM or DVD. School administrators and others in a school community could examine their school's performance on equity measures of

---

[1]Researchers can obtain access to the suppressed data on the Internet by signing an agreement that protects the confidentiality of data pertaining to individually identifiable students.

issues, such as school discipline; graduation rates; student retention; testing outcomes; access to advanced placement and gifted and talented programs; tendencies toward within-school segregation, particularly as related to tracking; and racial disparities in special education. The data for a given school could be compared with district, state, or national averages or to equity standards suggested by OCR. Significant deviations from the suggested "equity standards" could produce an interactive diagnostic exercise that would include descriptions of one or more research-based strategies or programs that could help the school to reduce the observed disparities and to achieve the suggested standards.

Public Reporting and Accountability

Another possible new use of E&S survey data is to incorporate it (and other information pertaining to access to learning resources) with the kinds of disaggregated achievement data that must be publicly reported under the No Child Left Behind Act. Under the act, schools are required to report achievement outcomes disaggregated by students' race and poverty status and for language minority students. Schools are held accountable for ensuring that students from each group make adequate progress toward achieving state-defined standards for learning.

Data related to access to learning opportunities and resources, such as those available from the E&S survey, should be an integral part of school data systems and should provide information that is helpful to diagnosing causes of disparities, whether or not they are related to discrimination. The inclusion by schools in routine public reporting of information on access to learning opportunities and resources, along with data on students' progress toward the achievement of standards, could be helpful in providing guidance about causes of disparities in outcomes as well as promising strategies to address them.

New Research Opportunities

Although the E&S survey is the most important source of information available on a wide range of topics related to access to learning opportunities and resources, data from the survey infrequently have been used in academic research. As discussed in Chapter 2, the survey is the sole national data source for a number of topics, including the application of disciplinary practices, classroom assignment data, and gender equity in sports. For many other issues, it is the only source for disaggregated data that can be linked to specific schools and districts or be projected to state and national levels.

The researchers who most frequently have made use of the data are graduate students working with education advocacy organizations and journalists (Peter McCabe, Office for Civil Rights, personal communication, 2002). These individuals typically conduct basic analyses—or, prior to web access, requested OCR staff to conduct basis analyses—documenting disparities that point to schools' failure to adequately serve students of diverse backgrounds. This kind of basic research is important and provides citizens with information that is essential to inform and motivate their involvement in school improvement efforts. However, using E&S data to conduct complex analyses to document the causes of disparities and identify the kinds of resources and educational practices that would be effective for overcoming them has been rare.

Several changes must take place if the data are to be more widely used in academic research. First, OCR must make the data more easily available to researchers. Neither OCR nor any other unit of the Department of Education routinely analyzes the data or publishes even basic findings from the survey in departmental publications. Outside the advocacy community, very few people are aware of the E&S survey—including many highly skilled and experienced education researchers. Some of the researchers contacted by the committee who were aware of the survey in the past have had difficulty

gaining access to usable datafiles. One researcher who works for an advocacy organization and who has used E&S survey data for more than 30 years told the committee, "These data are used to enforce civil rights laws. They are not intended for academic researchers" (Paul Smith, director of research for the Children's Defense Fund, personal communication, 2002).

Yet most of the researchers who worked with E&S survey data under the auspices of this committee found them to be of great interest. They could be an important resource on access to opportunities to learn if OCR could implement procedures to improve the quality of the data and make them available in a format that is amenable to research (see Chapter 4), although there currently does not appear to be any protocol for this. The Beyond 20/20 software that OCR provides to facilitate public access to E&S data over the Internet is well suited for displaying descriptive statistics or for doing cross-tabulations that can identify various kinds of disparities. However, for researchers, the software cannot be used to conduct more elaborate studies that might uncover the underlying causes for observed disparities.[2]

---

[2]After several months of discussions with the committee and internal work, OCR was able to provide the committee with files containing data from the 2000 survey in a format that lent itself to complex studies.

CONCLUSIONS

Disparities in educational outcomes and in learning opportunity among different types of students continue to be an important social problem. In that context, the committee has three principal conclusions about the E&S survey:

- The E&S survey, or an equivalent research instrument, continues to be needed to gather disaggregated data related to the equality of access to learning opportunities and resources that are traceable to individual classrooms, schools, and districts.
- The survey, although useful for civil rights enforcement, informing educational policy, and the conduct of research, is greatly underused.
- The survey can be made more useful by improvements to the content, the manner in which the survey is administered, and access to the valuable data it provides.

RECOMMENDATIONS

The committee offers recommendations in four categories: survey administration, improving data quality, in-

creasing access to the data, and disseminating survey findings. We end with an overall conclusion about the role of the E&S survey.

Survey Administration

- The mandated and certified collection of data related to possible violations of students' educational civil rights should be sustained.
- The survey should be supported by line-item funding in the department's budget to ensure its ongoing support at a level that is consistent with its continued quality.
- Because of the survey's importance, the department should consider undertaking a thorough study of the survey aimed at ensuring that it deals appropriately and in sufficient depth with the problems of discovering possible restrictions on students' learning opportunities and, if possible, reducing the reporting burden on schools and school systems.
- The E&S survey content and protocols should be coordinated with those of other department surveys to ensure consistency of definitions and the complementarity of the data and to eliminate redundant questions.

- The various stakeholders in the E&S survey—such as OCR enforcement staff, student advocates, state and local educators, and researchers—should discuss and explore the advantages and disadvantages of less frequent but more comprehensive surveys. With respect to a comprehensive survey, all schools should be surveyed, at a minimum, every 10 years, as was done in 2000.

Improving Data Quality

- Survey items should be revised to provide more useful and complete information on five topics:
  1. the qualifications and experience of teachers;
  2. the assignment of students to different types of classrooms and educational settings;
  3. the consequences for students of high-stakes testing;
  4. high school completion; and
  5. interscholastic athletics.

- OCR should ensure that respondents understand how to complete the survey accurately and thoroughly.
- OCR should carefully scrutinize the data that are collected for thoroughness and reliability.

Increasing Access to the Data

There are several steps that OCR should take to increase access to the E&S survey data:

- train staff to make more effective use of the survey data;
- continue to improve the software provided for public access to E&S survey data over the Internet;
- sponsor or support programs to train advocates, researchers, and educators to use the data for various purposes;
- make well-edited data available to researchers and others in a usable format, and provide a data manual and technical assistance;
- consider developing a small grants program to encourage research on the topic of access to learning opportunities using E&S survey data; and
- archive and preserve data from all surveys in a common format and make them accessible to researchers and other interested parties on disk or over the Internet, both for historical purposes and to enable researchers to track longitudinal trends.

Disseminating Survey Findings

Three steps should be taken by OCR to improve dissemination of E&S survey data:

- conduct or sponsor the conduct of basic tabulations of the data;
- include findings from analyses of the data in OCR's regular reports to Congress; and
- publicize the basic findings from the survey in widely disseminated government publications.

CONCLUSION

If the E&S survey were to be recreated today, with a core objective of providing information on equality of access to the kinds of learning opportunities and resources that would be useful in shaping education policy and practices and for informing efforts to protect students' civil rights, the issues addressed by the current survey would be an essential part of the framework. However, much more information also would be needed.

Knowledge of how the learning environment, peers, learning resources, teacher preparation, and curriculum affect student learning is constantly evolving, as is knowledge of how to measure variables associated with each of these categories of learning opportunities and resources. To admit that there is still much that is unknown about how to measure learning opportunities and resources should be a spur for more work. The committee urges the Department of Education not only to continue collecting the kind of information that currently is on the E&S survey, but also to constantly reassess its quality and utility. Finally, we urge the department to recommit itself to using this information not only to protect students' legally defined civil rights, but also to ensure that all students who are being held accountable for achieving high standards have equal access to the opportunities and resources needed to do so.

References and Bibliography

Abt Associates. (1993). *Prospects: The congressionally mandated study of educational growth and opportunity.* Washington, DC: U.S. Department of Education.

Adelman, C. (1999). *Answers in the toolbox: Academic intensity attendance patterns, and bachelor's degree attainment.* Washington, DC: U.S. Department of Education.

The Advancement Project and The Civil Rights Project. (2000). *Opportunities suspended: The devastating consequences of zero tolerance and school discipline policies.* Cambridge, MA: Harvard University, The Civil Rights Project.

American Educational Research Association, American Psychological Association, and The National Council on Measurement in Education. (1999). *Standards for educational and psychological testing.* Washington, DC: American Psychological Association.

American Federation of Teachers. (2001). *Making standards matter, 2001.* Washington, DC: Author.

Babcock, P.S. (2002). *Achievement disrupted: An exploration of achievement indicators and student performance.* Paper presented at the National Research Council meeting of the Committee on Improving Measures of Access to Equal Educational Opportunity, August 12–13, Woods Hole, MA.

Balfanz, R., McPartland, J., and Shaw, A. (2002). *Reconceptualizing extra help for high school students in a high standards era* (Report No. ED-99-CO-0160). Washington, DC: U.S. Department of Education, Office of Vocational and Adult Education.

Braddock, J.H., Sokol-Katz, J., and Greene A. (2002). *Uneven playing fields: State variations in boy's and girl's access to and participation in high school interscholastic sports.* Paper presented at the National Research Council meeting of the Committee on Improving Measures of Access to Equal Educational Opportunity, August 12–13, Woods Hole, MA.

Carey, N., Rowland, C., and Farris, E. (1998). *State survey on racial and ethnic classifications* (NCES 98-034). Washington, DC: U.S. Department of Education, Office of Educational Research and Improvement.

Carmona, L.A., Wheelock, M.A., and First, J. (1998). *A gathering storm: How Palm Beach County schools fail poor and minority children.* Boston: National Coalition of Advocates for Students.

Clark, K.B. (1963). *Prejudice and your child, 2nd edition.* Boston: Beacon Press.

Coleman, J. (1975). *Trends in school segregation, 1968–1973* (Paper #722-03-01 ed.). Washington, DC: The Urban Institute.

Council of Chief State School Officers. (2000). *Executive summary of state student assessment programs.* Washington, DC: Author.

Council of Economic Advisers. (2000). *Economic report of the President transmitted to the Congress, together with the annual report of the Council of Economic Advisers.* Washington, DC: U.S. Government Printing Office.

Croninger, R., and Douglas, K. (2002). *The prevalence of high-stakes testing in U.S. public elementary and secondary schools: Consequences for children of color.* Paper presented at the

National Research Council meeting of the Committee on Improving Measures of Access to Equal Educational Opportunity, August 12–13, Woods Hole, MA.

Cross, B. (2002). A time for action. In T. Johnson, J.E. Boyden, and W.J. Pittz (Eds.), *Racial profiling and punishment in U.S. public schools. How zero tolerance policies and high-stakes testing subvert academic excellence and racial equity* (pp. 5–8). Oakland: Applied Research Center.

Cuban, L. (1998). How schools change reforms: Redefining reform success and failure. *Teachers College Record, 99,* 453–477.

Darling-Hammond, L. (1997a). *Doing what matters most: Investing in quality teaching.* New York: National Commission on Teaching and America's Future.

Darling-Hammond, L. (1997b). *The right to learn.* San Francisco: Jossey-Bass.

DeFrancis, M. (1998). Where are they now: Many of the civil rights activists as well as those opposed to desegregation are still active in public affairs. *FOCUS* (August).

Editorial Projects in Education. (2001). *Quality counts, 2001.* Bethesda: Author.

Edley, C. (2002). Education reform in context: Research, politics and civil rights. In National Research Council, T. Ready, C. Edley, and C. Snow (Eds.), Division of Behavioral and Social Sciences and Education, *Achieving high educational standards for all* (pp. 123–148). Washington, DC: National Academy Press.

Farkas, G. (2002). *Racial disparities and discrimination in education: What do we know, how do we know it, and what do we need to know?* Paper presented at the National Research Council Workshop on Measuring Disparities in Education, July 1, Washington, DC.

Farley, R. (1975). Racial integration in public schools, 1967–1972: Assessing the effects of government policies. *Sociological Focus, 8*(1), 3–26.

Farley, R. (1976). Is Coleman right? *Social Policy, 6*(4), 1–10.

Farley, R. (1978). School integration in the United States. In F.D. Bean and W.F. Parker (Eds.), *The demography of racial and ethnic groups* (pp. 15–50). New York: Academic Press.

Farley, R. (1979). *School desegregation and enrollments in the nation's largest cities: An analysis of recent trends.* Paper presented at the 139th Annual Meeting of the American Statistical Association, August 13–16, Washington, DC.

Farley, R., and Taeuber, A.F. (1974). Racial segregation in the public schools. *American Journal of Sociology, 79*(4), 888–905.

Farley, R., Richard, T., and Wurdock, C. (1980). School desegregation and white flight: An investigation of competing models and their discrepant findings. *Sociology of Education, 53,* 123–139.

Ferguson, R. (2000). Certification test scores, teacher quality and student achievement. In D.W. Grissmer and J.M. Ross (Eds.), *Analytic issues in the assessment of student achievement.* Washington, DC: U.S. Department of Education, National Center for Education Statistics.

Ferguson, R., and Ladd, H. (1996). How and why money matters: An analysis of Alabama schools. In H. Ladd (Ed.), *Holding schools accountable: Performance-based reform in education* (pp. 265–298). Washington, DC: Brookings Institution.

Ferguson, R., and Mehta, J. (2002). Why racial integration and other policies since Brown v. Board of Education have only partially succeeded at narrowing the achievement gap. In National Research Council, T. Ready, C. Edley, and C. Snow (Eds.), Division of Behavioral and Social Sciences and Education, *Achieving high educational standards for all* (pp. 123–148). Washington, DC: National Academy Press.

Finn, J. (1982). Patterns in special education placement as revealed by the OCR surveys. In National Research Council, Panel on Selection and Placement of Students in Programs for the Mentally Retarded, Committee on Child Development Research and Public Policy, K.A. Heller, W.H. Holzman, and S. Messick (Eds.), *Placing children in special education: A strategy for equity* (pp. 322–381). Washington, DC: National Academy Press.

Ford, D.Y. (1998). The underrepresentation of minority students in gifted education: Problems and promises in recruitment and retention. *Journal of Special Education, 32,* 4–14.

Gallagher, E.D. (2002). *The odds ratios of failing MCAS among African Americans, Latinos, and whites and achieving at least proficiency on MCAS.* Available: http://www.es.umb.edu/edgwebp.htm#MCAS [August 12, 2002].

Gamaron, A. (1992). Access to excellence: Assignment to honors classes in the transition from middle to high school. *Educational Evaluation and Policy Analysis, 14*(3), 185–204.

Garcia Fierros, E., and Conroy, J.W. (2002). Double jeopardy: An exploration of restrictiveness and race in special education. In D.J. Losen and G. Orfield (Eds.), *Racial inequity in special education* (pp. 39–70). Cambridge, MA: Harvard Education Press.

Glennon, T. (2002). Evaluating the Office for Civil Rights' minority and special education project. In D.J. Losen and G. Orfield (Eds.), *Racial inequity in special education.* Cambridge, MA: Harvard Education Press.

Gordon, R. (1998). *Education and race.* Oakland: Applied Research Center.

Gordon, R., Della Piana, L., and Keleher, T. (2002). *Facing the consequences: An examination of racial discrimination in U.S. public schools.* Oakland: Applied Research Center.

Greenwald, R., Hedges, L.V., and Laine, R.D. (1996). The effect of school resources on student achievement. *Review of Educational Research, 66*(3), 361–396.

Gregg, M., and Leinhardt, G. (2002). Learning from the Birmingham Civil Rights Institute: Documenting teacher development. *American Educational Research Journal, 39,* 553–587.

Halpern, S.C. (1995). *On the limits of the law: The ironic legacy of Title VI of the 1964 Civil Rights Act.* Baltimore: Johns Hopkins University Press.

Halpern-Felsher, B.L., Connell, J.P., Spencer, M.B., Aber, J.L., Duncan, G.J., Clifford, E., Crinchlow, W.E., Usinger, P.A., Cole, S.P., LaRue, A., and Seidman, E. (1997). Neighborhood and family factors predicting educational risk and attainment in African American and white children and adolescents. In J. Brooks-Gunn, G.J. Duncan, and J.L. Aber (Eds.), *Neighborhood poverty: Context and consequences for children* (pp. 146–173). New York: Russell Sage Foundation.

Harry, B., and Anderson, M.G. (1995). The disproportionate placement of African American males in special education programs: A critique of the process. *Journal of Negro Education, 63*(4), 602–619.

Harry, B., Klingner, J.K., Sturges, K.M., and Moore, R.F. (2002). Of rocks and soft places: Using qualitative methods to investigate disproportionality. In D.J. Losen and G. Orfield (Eds.), *Racial inequality in special education* (pp. 71–92). Cambridge, MA: Harvard Education Press.

Hedges, L.V., Laine, R.D., and Greenwald, R. (1994). Does money matter? A meta-analysis of studies of the effects of differential inputs on student outcomes. *Educational Researcher, 23*(4), 5–14.

Hehir, T. (2002). IDEA and disproportionality: Federal enforcement, effective advocacy, and strategies for change. In D.J. Losen and G. Orfield (Eds.), *Racial inequity in special education* (pp. 219–238). Cambridge, MA: Harvard Education Press.

Hill, P.T., Campbell, C., and Harvey, J. (2000). *It takes a city: Getting serious about urban school reform.* Washington, DC: Brookings Institution.

Horn, C. (2002). *The intersection of race, class and English learner status.* Paper presented at the National Research Council meeting of the Committee on Improving Measures of Access to Equal Educational Opportunity, August 12–13, Woods Hole, MA.

Johnson, T., Emiko-Boyden, J., and Pittz, W.J. (2002). *Racial profiling and punishment in U.S. public schools: How zero tolerance policies and high stakes testing subvert academic excellence and racial equity.* Oakland, CA: Applied Research Center.

Kleiner, B., Porch, R., and Farris, E. (2002). *Public alternative schools and programs for students at risk of school failure: 2000–2001* (NCES 2002-004). Washington, DC: U.S. Department of Education, National Center for Education Statistics.

Kluger, R. (1976). *Simple justice: The history of Brown v. Board of Education and black America's struggle for equality.* New York: Knopf.

Lippman, L., Burns, S., and McArthur, E. (1996). *Urban schools: The challenge of location and poverty* (Report No. NCES 96-84). Washington, DC: U.S. Department of Education, Office of Educational Research and Improvement.

Losen, D.J., and Orfield, G. (2002a). Introduction. In D.J. Losen and G. Orfield (Eds.), *Racial inequality in special education*. Cambridge, MA: Harvard Education Press.

Losen, D.J., and Orfield, G. (Eds.). (2002b). *Racial inequality in special education*. Cambridge, MA: Harvard Education Press.

Losen, D.J., and Welner, K.G. (2002). Legal challenges to inappropriate and inadequate special education for minority children. In D.J. Losen and G. Orfield (Eds.), *Racial inequality in special education* (pp. 167–194). Cambridge, MA: Harvard Education Press.

MacMillan, D.L., and Reschly, D.J. (1998). Overrepresentation of minority students: The case for greater specificity or reconsideration of the variables examined. *Journal of Special Education, 32,* 15–24.

Mansfield, W., and Farris, W. (1992). *Office for Civil Rights survey redesign: A feasibility survey* (Report No. NCES 92-130). Washington, DC: U.S. Department of Education, Office of Educational Research and Improvement.

McFadden, A.C., Marsh, G.E., Price, B.J., and Hwang, Y. (1992). A study of race and gender bias in the punishment of school children. *Education and Treatment of Children, 15*(2), 140–146.

Mickelson, R.A. (2001). Subverting Swann: First-and-second-generation segregation in the Charlotte-Mecklenburg schools. *American Educational Research Journal,* August, (2), 215–252.

Minceberg, E.M., Cahn, N., Isaacson, M.R., and Lyons, J.J. (1989). Federal civil rights enforcement and elementary and secondary education since 1981. In R.C. Govan and W.L. Taylor (Eds.), *One nation, indivisible: The civil rights challenge for the 1990s. Report of the Citizens' Commission on Civil Rights* (pp. 88–127). Washington, DC: L&B.

Moses, R., and Cobb, C.E., Jr. (2001). *Radical equations: Math literacy and civil rights.* Boston: Beacon Press.

Myrdal, G. (1944). *An American dilemma.* New York: Harper and Row.

National Center for Education Statistics. (2001). *Digest of education statistics, 2000* (NCES 2001-034). Washington, DC: U.S. Department of Education.

National Coalition of Advocates for Students. (1998). Zero tolerance raises equity concerns across nation. *Mobilization for Equity* (November).

National Research Council. (1982). *Placing children in special education: A strategy for equity.* Panel on Selection and Placement of Students in Programs for the Mentally Retarded, Committee on Child Development Research and Public Policy, K.A. Heller, W.H. Holtzman, and S. Messick (Eds.), Commission on Behavioral and Social Sciences and Education. Washington, DC: National Academy Press.

National Research Council. (1989). *A common destiny: Blacks and American society.* Committee on the Status of Black Americans, G.D. Jaynes, and R.M. Williams, Commission on Behavioral and Social Sciences and Education. Washington, DC: National Academy Press.

National Research Council. (1999a). *Making money matter: Financing America's schools.* Committee on Education Finance, H.F. Ladd, and J.S. Hansen (Eds.), Commission on Behavioral and Social Sciences and Education. Washington, DC: National Academy Press.

National Research Council. (1999b). *Preventing reading difficulties in young children.* Committee on the Prevention of Reading Difficulties in Young Children, C.E. Snow, M.S. Burns, and P. Griffin (Eds.), Commission on Behavioral and Social Sciences and Education. Washington, DC: National Academy Press.

National Research Council. (1999c). *High stakes: Testing for tracking, promotion and graduation.* Committee on Appropriate Test Use, J.P. Heubert and J.M. Hauser (Eds.), Board on Testing and Assessment, Center for Education. Washington, DC: National Academy Press.

National Research Council. (2002a). *Achieving high educational standards for all: Conference summary.* T. Ready, C. Edley, and C.E. Snow (Eds.), Division of Behavioral and Social Sciences and Education. Washington, DC: National Academy Press.

National Research Council. (2002b). *Minority students in special and gifted education.* Committee on Minority Representation in Special Education, S.M. Donovan and C. Cross (Eds.), Division of Behavioral and Social Sciences and Education. Washington, DC: National Academy Press.

Neild, R.C., and Balfanz, R. (2001). *An extreme degree of difficulty: The educational demograph-*

ics of the ninth grade in Philadelphia. Baltimore: Johns Hopkins University, Center for the Social Organization of Schools.

Oakes, J. (1990). *Multiplying inequalities: The effects of race, social class and tracking on opportunities to learn mathematics and science.* Santa Monica, CA: RAND.

Orfield, G. (1969). *The reconstruction of Southern education: The schools and the 1964 Civil Rights Act.* New York: Wiley-Interscience.

Orfield, G. (1977). *Desegregation and the cities: The trends and the policy choices.* Washington, DC: U.S. Senate Committee on Human Resources.

Orfield, G. (1978). *Must we bus?* Washington, DC: Brookings Institution.

Orfield, G. (1986). *Public school desegregation in the United States, 1968–1980.* Washington, DC: Joint Center for Political Studies.

Orfield, G. (1996). *Dismantling desegregation: The quiet repeal of Brown v Board of Education.* New York: New Press.

Orfield, G. (2001a). *The 1964 Civil Rights Act and American education.* In B. Grofman (Ed.), *Legacies of the 1964 Civil Rights Act* (pp. 89–128). Charlottesville, VA: University of Virginia Press.

Orfield, G. (2001b). Why data collection matters. In W. Hutmacher, D. Cochrane, and N. Bottani (Eds.), *In pursuit of equity in education.* Boston: Kluwer Academic.

Orfield, G. (2001c). *Schools more separate: Consequences of a decade of resegregation.* Cambridge, MA: Harvard University, The Civil Rights Project.

Orfield, G., Bachmeier, D.J., and Eitle, T. (1997). Deepening segregation in American public schools. *Equity and Excellence in Education, 30*(2), 5–24.

Orfield, G., Schley, S., Glass, D., and Rardon, S. (1993). *The growth of segregation in American schools: Changing patterns of separation and poverty since 1968.* Alexandria, VA: National School Boards Association.

Orfield, G., and Yun, J. (1999). *Resegregation in American schools.* Cambridge, MA: Harvard University, The Civil Rights Project.

Osher, D., Woodruff, D., and Simms, A.E. (2002). Schools make a difference: The overrepresentation of African American youth in special education and the juvenile justice system. In D.J. Losen and G. Orfield (Eds.), *Racial equity*

in special education (pp. 93–116). Cambridge, MA: Harvard Education Press.

Oswald, D.P., Coutinho, M.J., Best, A.M., and Singh, N.N. (1999). Ethnic representation in special education: The influence of school-related economic and demographic variables. *Journal of Special Education, 32,* 194–206.

Panetta, L., and Gall, P. (1971). *Bring us together: The Nixon team and the civil rights retreat.* Philadelphia: J.B. Lippincott.

Patton, J.M. (1992). Assessment and identification of African-American learners with gifts and talents. *Exceptional Children, 59*(2), 150–159.

Puma, M., Karweit, N., Price, C., Ricciuti, A., Thompson, W., and Vaden-Kiernan, M. (1997). *Prospects: Student outcomes final report.* Washington, DC: U.S. Department of Education.

Rabekoff, E. (1990). Federal civil rights enforcement in elementary and secondary education. In S.M. Liss and W.L. Taylor (Eds.), *Lost opportunities: The civil rights record of the Bush administration mid-term: Report of the Citizens' Commission on Civil Rights* (pp. 37–51). Washington, DC: Citizens' Commission on Civil Rights.

Rabkin, J. (1980). Office for Civil Rights. In J.Q. Wilson (Ed.), *The politics of regulation* (pp. 304–353). New York: Basic Books.

Ready, D.D., and Lee, V.E. (2002). *Linking the Office for Civil Rights' elementary and secondary school survey with nationally representative educational datasets: Possibilities and problems.* Paper presented at the National Research Council Committee on Improving Measures of Access to Equal Educational Opportunity, August 12–13, Woods Hole, MA.

Rebell, M.A. (2002). Education adequacy, democracy, and the courts. In National Research Council, T. Ready, C. Edley, and C.E. Snow (Eds.), *Achieving high educational standards for all: Conference summary* (pp. 218–268). Washington, DC: National Academy Press.

Roderick, M., and Engel, M. (2001). The grasshopper and the ant: Motivational responses of low-achieving students to high stakes testing. *Educational Evaluation and Policy Analysis, 23,* 197–227.

Ross, S.L., and Yinger, J. (2002). *Detecting discrimination: A comparison of the methods* used by scholars and civil rights enforcement officials. Paper presented at the National Research Council Committee on Methods for Assessing Discrimination, July 1–3, Washington, DC.

Ryan, J.E. (2002). *What constitutes race discrimination in education? A legal perspective.* Paper presented at the National Research Council Committee on Methods for Assessing Discrimination, July 1–3, Washington, DC.

Sanders, W.L., and Horn, S.P. (1995). The Tennessee Value-Added Assessment System (TVAA): Mixed model methodology in educational assessment. In A.J. Shinkfield and D.L. Stufflebeam (Eds.), *Teacher evaluation: Guide to effective practice.* Boston: Kluwer Academic.

Sanders, W.L., and Rivers, J.C. (1996). *Cumulative and residual effects of teachers on future academic achievement.* Knoxville: University of Tennessee Value-Added Research and Assessment Center.

Shaw, S.R., and Braden, J.P. (1990). Race and gender bias in the administration of corporal punishment. *School Psychology Review, 19*(3), 378–383.

Slavin, R.E., Karweit, N.L., Wasik, B.A., Madden, N.A., and Dolan, L.J. (1994). Success for all: A comprehensive approach to prevention and early intervention. In R.E. Slavin, N.L. Karweit, and B.A. Wasik (Eds.), *Preventing early school failure* (pp. 175–205). Boston: Allyn and Bacon.

Sotomayor, F. (1974). *Para los ninos=For the children: Improving education for Mexican Americans.* Washington, DC: U.S. Commission on Civil Rights.

Swanson, H.L. (1999). *Intervention research for adolescents with learning disabilities: A meta-analysis of outcomes related to high-order processing.* Washington, DC: U.S. Department of Education.

Taeuber, K. (1990). Desegregation of public school districts: Persistence and change. *Phi Delta Kappan,* 18–24.

Taylor, W. (1971). Federal civil rights laws: Can they be made to work? *The George Washington Law Review, 39*(5), 971–1007.

U.S. Commission on Civil Rights. (1969). *Federal enforcement of school desegregation.* Washington, DC: Author.

U.S. Department of Education, Office for Civil Rights. (1999). *Ensuring access to equal educational opportunity.* Available http://www.ed.gov/offices/OCR/docs/ensure99.html [August 14, 2002].

U.S. Department of Education, Office for Civil Rights. (2000a). *Fiscal year 2000 annual report to Congress: Guaranteeing equal access to high standards education.* Washington, DC: Author.

U.S. Department of Education, Office for Civil Rights (2000b). *Fall 1998 elementary and secondary civil rights compliance report: Time series documentation.* Washington, DC: Author.

U.S. Department of Health Education and Welfare, Office for Civil Rights. (1978). *Directory of elementary and secondary school districts, and schools in selected school districts: School year 1976–1977, volumes I and II.* Washington, DC: Author.

Vaughn, S., Gersten, R., and Chard, D.J. (2000). The underlying message in LD intervention research: Findings from research syntheses. *Exceptional Children, 67*(1), 99–114.

Welch, F. (1987). A reconsideration of the impact of school desegregation programs on public school enrollment of white students, 1968–1976. *Sociology of Education, 60*(October), 215–221.

WESTAT. (1997). *Redesign considerations: Elementary and secondary school civil rights compliance report* (Report No. EB950390010). Washington, DC: U.S. Department of Education.

Appendix A

Synopses of Papers
Prepared for the Committee

This appendix contains synopses of five papers commissioned by the committee and presented at its second meeting, August 12, 2002, at Woods Hole, Massachusetts. The committee's purpose in commissioning these papers was to obtain information about the value of E&S survey data for research on current educational policies and practices. The committee considered the findings of these papers, as well as the researchers' accounts of their experiences in working with the E&S data, in its deliberations.

Given the short amount of time available to the researchers, the papers should be considered preliminary analyses; they address the following topics:

- the effects of school disciplinary policies,
- gender equity in sports,
- the impact of high-stakes testing,

- the degree of segregation of English-language learners, and
- the availability of services for English-language learners.

The committee encouraged the researchers to continue the analyses begun under the auspices of the committee and to independently publish papers based on their analyses, as appropriate.

ACHIEVEMENT DISRUPTED

Philip Babcock
University of California, San Diego

This paper investigates correlations between test scores and school expulsion rates to determine to what degree these correlations might be driven by policy. The investigation is framed as part of a larger inquiry into the underlying causes of peer effects.

The analysis begins with a basic description of the data, e.g., histograms of expulsions and suspensions, sample covariances between test scores and demographic measures. The main body of the paper explores implications of a specific theory of peer effects.

While studies suggest that peer effects have an impact on measures of student performance, the specific mechanism by which they operate remains unclear. One explanation is that disruptive behavior by a student impedes the learning of every other student in a classroom. If so, then systematic efforts made by teachers and educators to reduce disruptive behavior ought to lead to higher test scores, other things being equal. The paper examines an implication of this basic theory. Taking suspension rates and expulsion rates as proxies for discipline, one could attempt to determine the effect of discipline on test scores by means of a simple ordinary least squares regression. The immediate difficulty with such an approach is that there is apt to be correlation between the error term and the regressors. One would expect higher rates of expulsion at schools whose students come from troubled or dysfunctional social environments and have behavior problems because of unobserved factors. These students might perform poorly on tests because of the same unobserved traits that affect the rate of expulsion.

The paper attempts to distinguish between the two sources of correlation, endogenous and exogenous, by constructing a panel of test scores, suspension rates, and expulsion rates for ninth graders in California, school by school, using data from 1998 and 2000. The analysis includes both fixed-effect and between-effects estimates of the coefficients on expulsion and suspension, the assumption being that neighborhood effects for a given school changed little in two years.

The analysis indicates that when one does not control for school-specific unobserved effects, higher rates of expulsion are associated with lower math scores. The paper argues that this is evidence of endogeneity: schools whose students have been unobservably disadvantaged by their local environment exhibit more behavior problems and also lower math scores. In the fixed-effects regression, however, the correlation is positive. Holding constant the school, an increase in expulsion rates between 1998 and 2000 was associated with an increase in math scores. The strictness of discipline policy, then, might have a positive effect on test scores. The result is merely suggestive, however: the analysis was not robust to heteroskedastic specifications or to reasonable alterations of the dataset.

The paper concludes with a discussion of the database of the Office for Civil Rights (OCR), including some suggestions for additions that would be helpful in this area of research.

UNEVEN PLAYING FIELDS: STATE VARIATIONS IN BOY'S AND GIRL'S ACCESS TO AND PARTICIPATION IN HIGH SCHOOL INTERSCHOLASTIC SPORTS

Jomills Henry Braddock II, Jan Sokol-Katz, and Anthony Greene
University of Miami

This year marks the 30th anniversary of Title IX. Yet despite considerable progress and the need for further improvement, Title IX is facing increased opposition (including numerous legal challenges) and scrutiny (Secretary of Education's Commission on Opportunities in Athletics). Using aggregate data from OCR, combined with demographic and contextual data available from the National Center for Educational Statistics, this paper (1) analyzes state-level disparities in boy's and girl's access to both single- and mixed-gender interscholastic athletic programs and their patterns of participation in single- and mixed-gender school sports and (2) examines demographic and contextual correlates of variation among states in relative access and participation of boys and girls in interscholastic athletics.

While unity, or even virtual parity, has been achieved in only a handful of states, we find wide variation among the states in how equitably girls and boys have access to both single- and mixed-gender interscholastic sports and teams. We also find that states vary widely in patterns of participation in single- and mixed-gender sports activities. Regression analyses suggest that variation in gender disparities in single-gender athletic participation opportunities (number of sports and teams offered to male and female students) among states can be predicted by a combination of contextual characteristics (median household income) and school demographics (percentage of white enrollments). And variation in gender disparities in single-gender athletic participation rates (ratio of male to female students participating in single-gender sports) among states can be strongly predicted by participation opportunities (number of sports and teams offered to male and female students) and educational investment (per pupil expenditures). This analysis also suggests that the impact on gender differences in participation rates of school demographics (percentage of white enrollments) is largely indirect and mediated by participation opportunities (number of sports and teams offered to male and female students). With regard to gender disparities in mixed-gender athletic participation opportunities (number of sports and teams offered to both male and female students) among states, our model is not adequate to explain state variations. However, gender disparities in mixed-gender athletic participation rates among states can be modestly predicted by a combination of contextual characteristics (median household income) and educational investment (per pupil expenditures).

Our analysis suggests that monitoring gender equity in athletic access and participation is crucial. Because the OCR compliance reports are the most reliable sources of data on this equity issue, it will be important to both continue and strengthen data collection efforts. Specifically, additional information is required to better understand the relationship between enrollment racial composition and access to and participation in single-gender athletics. Recent questions that have been raised about whether black females are benefiting from Title IX could be informed with better information about characteristics of student participants. Better information regarding specific sports offered is also important. This would help clarify general understanding of such issues as the types of sports available to girls and boys as both single- and mixed-gender activities. Next steps should include multilevel analyses taking into account variations at the school, district, and state levels.

PREVALENCE OF HIGH-STAKES TESTING IN U.S. PUBLIC ELEMENTARY AND SECONDARY SCHOOLS: CONSEQUENCES FOR MINORITY CHILDREN

Robert G. Croninger and Karen Douglas
University of Maryland

Recent policy initiatives have placed increasing importance on the implementation of high-stakes tests for reforming public elementary and secondary schools. Provisions in the 2001 Elementary and Secondary Education Act call for increased testing to determine if students and teachers are meeting high academic standards, as do new state-level policies that seek to align state testing practices with rigorous content and performance standards for students and teachers. Forty-six states have developed or are in the process developing testing policies aligned to grade-level content and performance standards (see National Research Council, 1999). Of these states, 18 require that students pass some form of an exit examination before receiving a high school diploma (see Council of Chief

State School Officers 2000), while 3 states require schools to use state standards and assessments in making promotional decisions for elementary school or middle school students. Four additional states plan to make promotion contingent on high-stakes tests by 2003 (see Editorial Projects in Education, 2001).

Although high-stakes testing provides new opportunities for holding individuals and schools accountable for the quality of educational opportunities that they provide students (see Weckstein, 1999), it also raises important equity issues about the actual consequences of testing for specific populations of students. If minority students and the schools that they attend disproportionately bear the burden of high-stakes testing both in terms of the requirements to pass high-stakes tests and the sanctions imposed for failure to do so, then the new wave of testing polices and practices may not promote more equitable educational opportunities. On the contrary, it may promote unfair (National Research Council, 1999) and perhaps illegal testing practices (*Debra P. v. Turlington,* 474 F. Supp. 244, M.D. Fla. 1979) that further deny minority students access to valuable educational opportunities (Howe, 1997). As policy makers at all levels of government call for new and more demanding testing practices, it is increasingly important

that we examine the consequences of such practices against not only standards of academic rigor but also standards of educational equity.

We know surprisingly little about either the prevalence or consequences of high-stakes testing given the attention it has received in recent education policies. While a number of surveys have been conducted to characterize high-stakes testing policies at the state level (e.g., see American Federation of Teachers, 2001; Council of Chief State School Officers, 2000; Editorial Projects in Education, 2001), only the 2000 OCR E&S survey provides data about the prevalence and consequences of high-stakes testing as practiced by individual districts and schools. Although the states have taken the lead in implementing high-stakes testing, there is ample evidence that individual districts (e.g., Chicago Public Schools) have implemented high-stakes testing policies independent of state legislators and education officials (National Research Council, 1999). Even in states that require the use of high-stakes tests for promotional decisions or the certification of high school graduation, there may be considerable variability in the implementation of policies or in the consequences of policies for students. Moreover, because the 2000 E&S survey asked schools to report the pass and failure rates of students by race, gender, and disability status, it may be possible to use these data to investigate not only the prevalence of high-stakes testing practices but also their immediate consequences for specific student populations. These types of data are unavailable elsewhere.

We propose to explore the feasibility of using the 2000 E&S survey data to analyze the prevalence and consequences of high-stakes testing practices in U.S. elementary and secondary schools. Specifically, we propose to (1) evaluate the research utility of the current format for E&S survey data by attempting to download data about testing (Tables 12a and 12b) using the Beyond 20/20 interface; (2) link these data to the 2000 Common Core of Data (CCD); and (3) examine the validity of E&S survey data by comparing these data with other data sources (e.g., OCR and CCD enrollment data; OCR and Council of Chief State School Officers reports of testing data by states).

The analytic method used will depend on the variability between schools, districts, and states in testing practices and their consequences. The research questions that we propose to address in the study include:

- What is the prevalence of high-stakes testing as a requirement for grade promotion? How do these practices vary by state, districts, and schools? Are there differences by race in who is and is not held to test-based promotional requirements?

- What are the immediate consequences of using high-stakes testing to make promotion decisions for students? Are there differences by race in who passes or fails these tests?

- What is the prevalence of high-stakes testing as a requirement for high school graduation? How do these practices vary by state, districts, and schools? Are there differences by race in who is and is not required to pass an exit examination before receiving a high school diploma?

- What are the immediate consequences of using high-stakes testing as a graduation requirement for students? Are there differences by race in who passes or fails these tests?

References

American Federation of Teachers. (2001). *Making standards matters 2001*. Washington, DC: Author.

Council of Chief State School Officers. (2000). *Executive summary of state student assessment programs*. Washington, DC: Author.

Editorial Projects in Education. (2001, January 11). Standards related policies. *Education Week*, 68–87.

Howe, K.R. (1997). *Understanding equal educational opportunity. Social justice, democracy, and schooling*. New York: Teachers College Press.

National Research Council. (1999). *High stakes: Testing for tracking, promotion, and graduation.* Committee on Appropriate Test Use, J.P. Heubert and R.M. Hauser (Eds.). Board on Testing and Assessment, Center for Education. Washington, DC: National Academy Press.

Weckstein, P. (1999). School reform and enforceable rights to quality education. In J.A. Heubert (Ed.), *Law and school reform. Six strategies for promoting educational equity* (pp. 306–389). New Haven: Yale.

THE INTERSECTION OF ENGLISH-LANGUAGE LEARNING, RACE, AND POVERTY

Catherine Horn
Harvard University

The school-age population in the United States is becoming increasingly diverse. As examples, in 1999, 20 percent of the school-age children had at least one foreign-born parent, including 5 percent of elementary and secondary students who were themselves foreign born (Jamieson, Curry, and Martinez, 2001). The number of school-age students who are Hispanic rose from 13 percent in 1993 to 17 percent nationally in 2000 (U.S. Census Bureau, 2001). Also in 2000, almost 1 in 10 public school students was an English-language learner, and Spanish continues to be the predominant language background of the students receiving English-language learner services (Kindler, 2002).

Two of the reported variables in the E&S survey broken out by race/ethnicity are the numbers of students needing and enrolled in programs for English-language learners. Coupled with data from the National Center for Educational Statistics' CCD, this information begins to shed light on the ways in which English proficiency, race, and income intersect. To that end, this paper looks closely at the following broad research question: In what ways do concentrations of English-language learners interact with poverty and race?

Studying the ways in which these three important characteristics intertwine is paramount to better understanding the influence of each on its own. For example, on an aggregate level, we know that while 18 percent of white, non-Hispanic students and 7 percent of black students have foreign-born parents, 88 and 65 percent of Asians and Hispanics, respectively, have at least one parent who was born outside the United States; 25 and 18 percent respectively were themselves foreign born (Jamieson, Curry, and Martinez, 2001). Of course, while these numbers do not exactly reflect the pool of students needing English-language services, they are certainly an indication that certain racial and ethnic groups may have a disproportionate need for them. Through the use of data included in the E&S survey and CCD, we can explore the racial and ethnic and socio-economic conditions of the schools in which large proportions of English-language learners are present, in comparison with schools with smaller proportions.

The paper first explores the interrelationship of English proficiency, race, and income by presenting descriptive tables, including, but not limited too, the following:

- by deciles, concentrations of English-language learners by concentrations of poverty;
- by deciles, concentrations of English-language learners by concentrations of nonwhite school demographics;
- concentrations of English-language learners by location (e.g., rural, suburban, urban); and
- districts with the highest percentage of their total student population needing or receiving English-language learner services.

The work then turns its focus toward those schools in which concentrations of English-language learners are the highest. Using a modified exposure index, the paper presents a series of findings displaying schools' racial distributions for the average student in a school with a high concentration of those students. So, for example, these data show the percentage of white students in schools with high concentrations of English-language learners attended by the typical black or Hispanic student. These findings are compared with similar exposure indices, by race, for schools in general. To the extent possible, this paper also explores the racial distributions for the average student in a school with high concentrations of poverty.

The paper concludes with a discussion of the policy implications of the findings. From a civil rights perspective, such information should be useful, for example, in considering the confounding impacts of legal decisions to end court-mandated desegregation. The paper also includes a brief discussion about the viability and limitations of the E&S survey data for continued research around the issues of English-language learners, race, and poverty and suggestions for better data collection to understand these issues.

References

Jamieson, A., Curry, A., and Martinez, G. (2001). *School enrollment in the United States—Social and economic characteristics of students, October, 1999.* Current Population Reports. Washington, DC: U.S. Census Bureau. Available: http://www.census.gov/population/www/socdemo/school.html.

Kindler, A. (May, 2002). *Survey of the states' limited English proficient students and available educational programs and services 1999–2000 summary report.* Washington, DC: George Washington University, The National Clearinghouse for English-Language Acquisition and

Language Instruction Educational Programs. Available: http://www.ncela.gwu.edu/ncbepubs/seareports/99-00/sea9900.pdf.

U.S. Census Bureau. (2001). *Table A-1: School enrollment of the population 3 to 34 years old, by level and control of school, race, and Hispanic origin: October 1955 to 2000.* Washington, DC: Author. Available: http://www.census.gov/population/www/socdemo/school.html.

LINKING THE OFFICE FOR CIVIL RIGHTS' ELEMENTARY AND SECONDARY SCHOOL SURVEY WITH NATIONALLY REPRESENTATIVE EDUCATIONAL DATASETS: POSSIBILITIES AND PROBLEMS

Douglas D. Ready and Valerie E. Lee
University of Michigan

We began investigating the E&S survey data with two objectives in mind. First, we sought to determine the feasibility of linking the E&S to data from the Early Childhood Longitudinal Study (ECLS-K), also collected by the U.S. Department of Education.[1] Second, if we could indeed link the datasets, we intended to perform analyses possible only through the combined E&S survey and ECLS-K datasets. Our hope was that E&S survey data would provide school-level measures that were useful to such analyses and unavailable on ECLS-K. Conversely, because the E&S survey includes no student-level outcomes, we hoped to augment those data with student-level social and academic measures available on ECLS-K

After considerable effort, we were successful in attaining our first goal; it is indeed possible to create a combined E&S survey and ECLS-K datafile. The standard NCES 12-digit school identification codes, which are included on both E&S survey and ECLS-K data (restricted file only), make this linkage possible.[2] Using these codes to match E&S survey to ECLS-K schools, we ultimately linked 687 public schools.[3]

Our next task was to identify important questions that could be answered

---

[1] Both of the authors of this paper are engaged in a multiyear study using ECLS-K data that is funded by the U.S. Department of Education.

[2] We accomplished this by saving the E&S survey data as a comma-separated file, opening the file in SPSS, and then merging the file with the ECLS-K data using the NCES school ID common to both files.

[3] We used the 2000 E&S survey data. ECLS-K kindergarten data were collected during the 1998–1999 school year with first grade data collected the following school year. The school sample includes only public schools (as E&S contains only public schools) that have both kindergarten and first grade and that enroll at least five ECLS-K students. Twenty ECLS-K schools that matched these criteria were not located in the E&S survey dataset. The student sample includes children with fall-K, spring-K, and spring-first test scores; full data on race, socioeconomic status, and gender; and were in the same school for kindergarten and first grade. The resulting average within-school ECLS-K sample size is 15.03.

only through analyses using this newly created dataset. Such analyses are by nature multilevel, requiring the use of hierarchical linear modeling. Our goal was to investigate how school-level characteristics (provided by the E&S survey) influence student outcomes (provided by ECLS-K), and whether those characteristics influence the *relationship* between student characteristics and student outcomes. For example, previous research suggests that the proportion of schools' minority enrollment is related to student achievement, even after accounting for the average social class and prior achievement of the students they enroll. Although research on racial segregation is personally and professionally important to us, we realized that ECLS-K already contains information about each school's racial composition. The E&S survey also contains important information about teacher certification, but again, this information is included in ECLS-K. Because ECLS-K studies children, classrooms, and schools in the early elementary grades (at present, kindergarten and first grade), the E&S survey high school measures were not salient. The E&S survey also contains several measures pertaining to student suspension, expulsion, and corporal punishment, which are again not particularly relevant in schools attended by kindergartners and first graders.

We settled on a series of research questions involving access to limited-English-proficiency (LEP) programs in public elementary schools that offer kindergarten. We were interested in whether first grade LEP students suffered academically by attending schools in which access to LEP services was limited or restricted. The E&S survey contains a measure indicating the number of students in each school *needing* LEP services and another indicating the number that actually *receive* LEP services. Our next step would have been to create a measure for each school indicating the extent to which LEP students were being denied access to LEP services—the proportion eligible for but not receiving LEP services. However, initial analyses revealed the correlation between the two measures was: 96. This is welcome news for LEP students and for researchers interested in equity; students who need LEP services in U.S. public schools generally receive them, at least according to the school staff who filled out the survey. For us, however, the fact that these two measures are essentially identical precluded their use in our investigation because of the virtual lack of variability between measures.

Because of the lack of informative school-level measures and the dearth of variables unique to the E&S survey, we decided we could not conduct empirical analyses using the combined E&S/ ECLS-K datafile. We therefore altered both our approach and our questions substantially. Instead of restricting our focus to one or two empirical questions, we broadened our efforts to investigate the general utility of the E&S survey data.

In the first section of the full report, we document the extent to which the E&S survey includes data in common with three other large, widely used datasets collected by the U.S. Department of Education: the CCD, the National Educational Longitudinal Study (NELS:88), and the ECLS-K. Of course, the timing of all data collections must coincide. For example, students in NELS:88 began high school in 1988 and graduated (most of them) in 1992. Our findings are displayed in a series of tables that indicate which E&S survey measures are redundant and available on other datasets and which data are unique to E&S survey. In a concluding section of the report we share our views regarding the value of the E&S survey from the standpoint we know best: as quantitative researchers interested in studying educational equity using large, nationally representative databases.

Appendix B

Overview of Findings from the 2000 E&S Survey

This appendix contains 21 figures that show basic findings from the 2000 E&S survey. They depict racial, ethnic, and gender differences for many of the variables in the survey. Because the 2000 E&S survey was administered to nearly 100 percent of public schools in the United States for the first time since 1976, the figures are included to provide the reader with a pictorial overview of the data.

Most of the figures are based solely on data from the E&S survey. Figures depicting relationships between the average socioeconomic status of the students of a school[1] and that school's composition by race, ethnicity, and English proficiency also rely on data from the Common Core of Data that were linked to the 2000 E&S dataset.

The figures were produced for the committee by Douglas E. Ready of the University of Michigan.

---

[1] The measure of the average socioeconomic status of a school is based on the percentage of students enrolled who are eligible for free or reduced-price lunch.

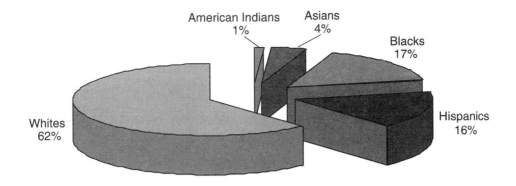

FIGURE B-1 Race and ethnicity of U.S. public school students, 2000–2001. (Data are projections from the 2000 E&S survey.)

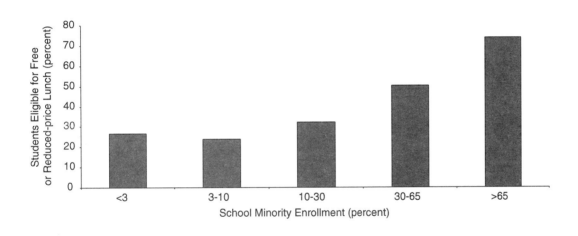

FIGURE B-2 School minority enrollment and poverty concentration. (Minority enrollment data from the 2000 E&S survey. Data on free/reduced-price lunch eligibility from the Common Core of Data, 2000–2001. The data do not include special education and alternative schools [school n = 69,029]. Minority enrollment includes American Indians, blacks, and Hispanics. Data on free/reduced-price lunch eligibility are not available for the following states: AZ, CT, IL, TN and WA.)

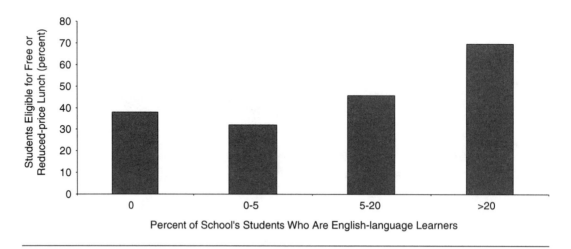

FIGURE B-3 Poverty concentration and English-language learners. (English-language learner data from the 2000 E&S survey. Data on free/reduced-price lunch eligibility from Common Core of Data, 2000–2001. Data on free/reduced-price lunch eligibility are not available for the following states: AZ, CT, IL, TN and WA.)

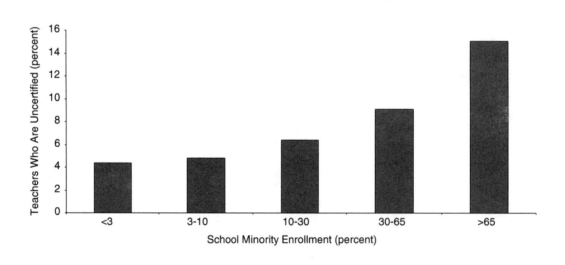

FIGURE B-4 School minority enrollment and uncertified teachers.
*9 percent of U.S. teachers do not hold state teaching certificates. (Data from the 2000 E&S survey. The data do not include special education and alternative schools [school n = 81,858]. Minority enrollment includes American Indians, blacks, and Hispanics.)

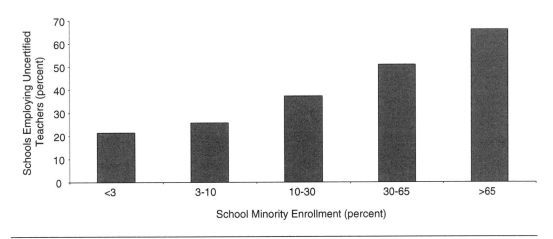

FIGURE B-5 Minority enrollment and schools employing any uncertified teachers.
*60 percent of all public schools employ only state-certified teachers. (Data from the 2000 E&S survey. The data do not include special education or alternative schools [school n = 82,341]. Minority enrollment includes American Indians, blacks, and Hispanics.)

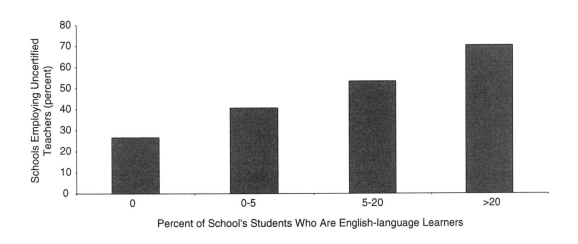

FIGURE B-6 Uncertified teachers and English-language learners. (Data from the E&S survey. The data do not include special education and alternative schools [school n = 80, 424].)

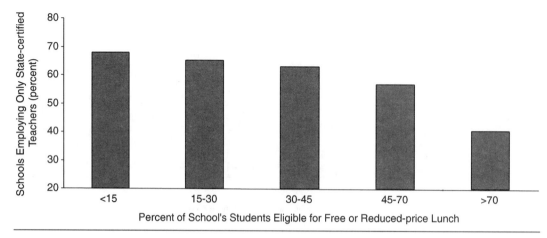

FIGURE B-7 Poverty concentration and access to state-certified teachers. (Teacher certification data from the 2000 E&S survey. Data on free/reduced-price lunch eligibility from Common Core of Data, 2000–2001. The data do not include special education and alternative schools [school n = 64,969]. Data on free/reduced-price lunch eligibility are not available for the following states: AZ, CT, IL, TN, and WA.)

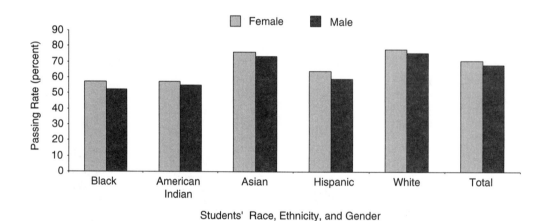

FIGURE B-8 Proportion of high school students passing tests that are the sole criterion for graduation. (Data from the 2000 E&S survey. Schools were instructed to consider a test the "sole criterion" for graduation if "all students were required to take a district-or-state-required test, and must pass the test to graduate from high school." A total of 2,652 schools required these tests; 585,709 high school students took such tests, 406,502 of whom passed.)

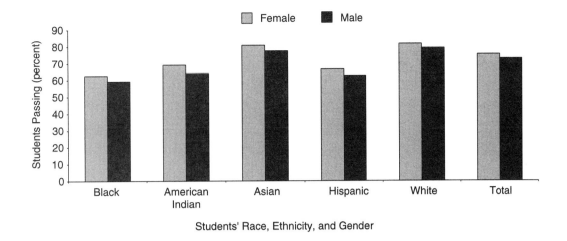

FIGURE B-9 Proportion of high school students passing tests that are a significant criterion for graduation. (Data from the 2000 E&S survey. Schools were instructed to consider a test of "significant criterion for graduation" if all students were required to take a district-or-state-required test, and the test is an important criterion in the decision on whether or not the student graduates from high school, but other criteria, such as teacher recommendations or the student's grades were used in the graduation decision." A total of 5,269 high schools required these tests; 1,149,780 high school students took such tests, 853,625 of whom passed.)

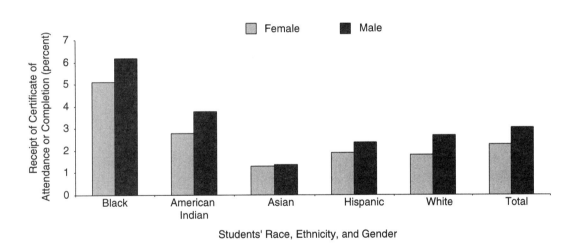

FIGURE B-10 Proportion of U.S. high school graduates receiving certificates of attendance or completion. (Data are projections from the 2000 E&S survey. Projections by the Office for Civil Rights (OCR) indicate that there were 2,605,843 public high school graduates in 2000, of whom 69,081 received certificates of attendance or completion. OCR defines a certificate of attendance or completion as "an award of less than a regular diploma, or a modified diploma, or fulfillment of an IEP for students with disabilities.")

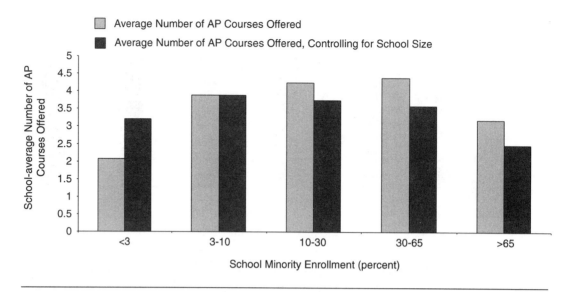

FIGURE B-11 High school minority enrollment and access to advanced placement (AP) courses. (Data from the 2000 E&S survey. The data do not include special education and alternative high schools [school n = 15,920]. Minority enrollment includes American Indians, blacks, and Hispanics.)

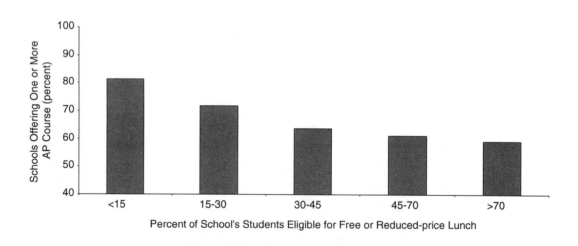

FIGURE B-12 Poverty concentration and access to advanced placement (AP) courses. (AP data from the 2000 E&S survey. Data on free/reduced-price lunch eligibility from CCD, 2000–2001. The data do not include special education or alternative high schools [school n = 11,299]. "High schools" include all schools enrolling eleventh and twelfth graders. Data on free/reduced-price lunch eligibility not available for the following states: AZ, CT, IL, TN, and WA.)

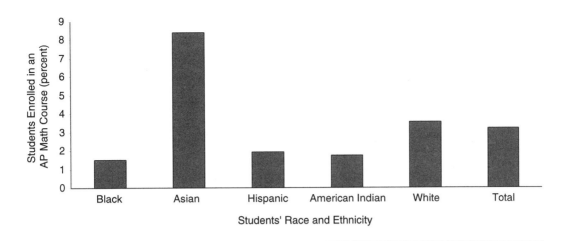

FIGURE B-13 Proportion of high school students enrolled in an advanced placement (AP) math course. (Data from the 2000 E&S survey. Data include only students attending high schools in which AP math courses are available. Alternative and special education high schools are not included in these calculations.)

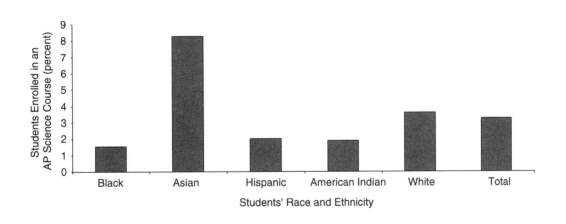

FIGURE B-14 Proportion of high school students enrolled in an advanced placement (AP) science course. (Data from the 2000 E&S survey. Data include only students attending high schools in which AP science courses are available. Alternative and special education high schools are not included in these calculations.)

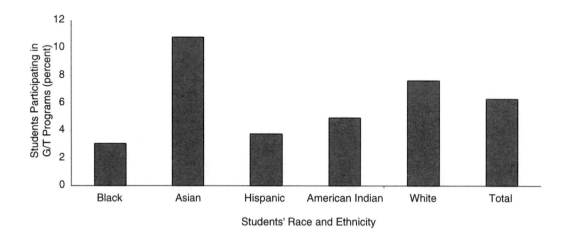

FIGURE B-15 Proportion of students participating in gifted and talented programs. (Data from the 2000 E&S survey. Gifted and talented programs are defined by OCR as special programs during regular school hours for students who possess unusually high academic ability or a specialized talent or aptitude, such as in literature or the arts.)

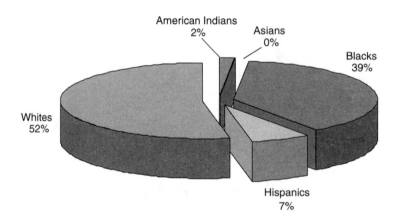

FIGURE B-16 Racial and ethnic composition of students receiving corporal punishment (n = 342,031). (Projections from the 2000 E&S survey. Corporal punishment includes paddling, spanking, and other forms of physical punishment. Students were counted only once, regardless of the number of times they received corporal punishment.)

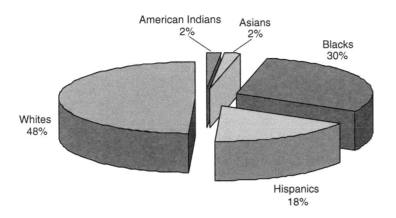

FIGURE B-17 Racial and ethnic composition of expelled students. (Projections from the 2000 E&S survey. Expulsion is defined as the exclusion from school for disciplinary reasons that results in the student's removal from school attendance rolls or that meets the criteria for expulsion as defined by the appropriate state or local school authority.)

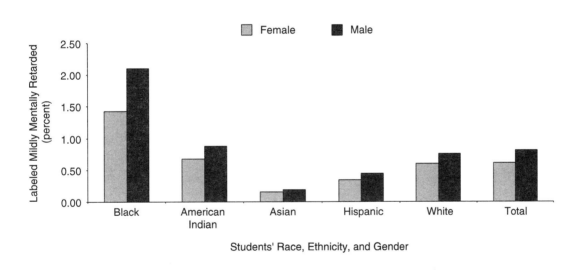

FIGURE B-18 Proportion of U.S. public school students labeled mildly mentally retarded by race, ethnicity, and gender. (Data from the 2000 E&S survey. Total student n = 45,837, 331; E&S data indicate 327,397 students were labeled mildly mentally retarded.

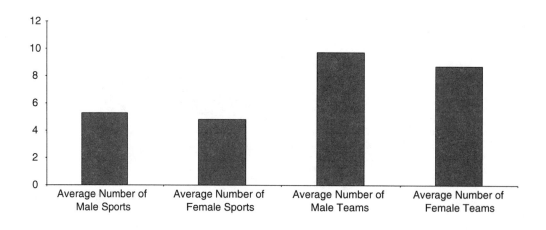

FIGURE B-19 Gender equity in high school athletics. (Data from the 2000 E&S survey. The data do not include special education and alternative high schools [school n = 15,216].)

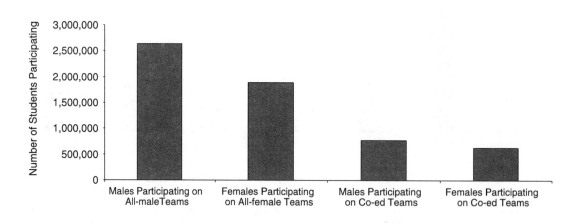

FIGURE B-20 Gender and participation in high school athletics. (Data from the 2000 E&S survey. The data do not include special education and alternative high schools [school n = 14,216].)

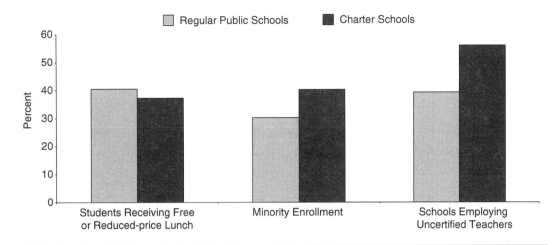

FIGURE B-21 Comparing charter and regular public schools: Student and teacher characteristics. (Data from the 2000 E&S survey. The data do not include special education and alternative schools. Data on free/reduced-price lunch eligibility from Common Core of Data, 2000–2001. Minority enrollment includes American Indians, blacks, and Hispanics. For free/reduced-price lunch comparison, regular school n = 68,424; charter school n = 573. For other comparisons, regular school n = 81, 297; charter school n = 1,007.)

Appendix C

2000 E&S Survey

2000 ELEMENTARY AND SECONDARY SCHOOL
CIVIL RIGHTS COMPLIANCE REPORT
DISTRICT SUMMARY: ED101

U.S. Department of Education, Office for Civil Rights
Washington, D.C. 20202-1172
Due Date: February 23, 2001

REPORTING REQUIREMENT

This Compliance Report is required by the U.S. Department of Education under Title VI of the Civil Rights Act of 1964, Title IX of the Education Amendments of 1972, and Section 504 of the Rehabilitation Act of 1973. Implementing Regulations are issued to carry out the purposes of Title VI of the Civil Rights Act of 1964, at 34 CFR 100.6(b); Title IX Regulations are at 34 CFR 106.71; and Section 504 Regulations are at 34 CFR 104.61.

Public Burden Statement. According to the Paperwork Reduction Act of 1995, no persons are required to respond to a collection of information unless such collection displays a valid OMB control number. The valid OMB control number for this information collection is 1870-0500. The time required to complete this information collection is estimated to average 7.5 hours per response, including the time to review instructions, research existing data resources, gather the data needed, and complete and review the information collection. **If you have any comments concerning the accuracy of the time estimate(s) or suggestions for improving this form, please write to U.S. Department of Education, Washington, D.C. 20202-1172. If you have comments or concerns regarding the status of your individual submission of this form, write directly to:** Compliance Reports Coordinator, U.S. Department of Education, 400 Maryland Avenue S.W., Room 5316, Switzer Building, Washington, D.C. 20202-1172.

GENERAL INSTRUCTIONS

- Please print legibly using a black ball-point pen.

- If you select to report via a mainframe computer cartridge or PC diskette, please see separate documentation.

- Information should be reported as of October 1, 2000, or the nearest convenient date prior to December 15, 2000, unless otherwise noted. Whenever possible, information should be provided consistent with the date of the special education Child Count in your state.

- **In order to allow us to distinguish between no students for a given item and non-applicability of that item for your district, please enter a response.** If the answer to a given item is *NONE*, enter *zero (0)* in the appropriate space. If a particular item is not applicable in your case, enter an X in the far right column.

- The certification signature block must be completed for the district by the superintendent or authorized representative. The certification pertains to *all* items on the ED-101 and ED-102 forms.

- **Please mail original forms to the Office for Civil Rights.** Retain photocopied ED-101 and ED-102 forms for your reference for two years from the date signed. The mailing address is: Compliance Report Project Office, Office for Civil Rights, 400 Maryland Avenue, S.W., Washington, D.C. 20202-1172.

- Please add the telephone number and the **FAX telephone number** in the appropriate boxes on the form for the individual in your district who can respond to questions we might have.

Page 1: ED101 - Instruction Sheet

DEFINITIONS

PUBLIC SCHOOL. An institution that provides pre-school, elementary, and/or secondary instruction; has one or more grade groupings *(pre-kindergarten through 12)* or is ungraded; has one or more teachers to give instruction; is located in one or more buildings, has an assigned administrator(s); receives public funds as its primary support; and is operated by an education agency. Public schools include charter schools that receive public funding from local or state sources.

PUBLIC SCHOOL MEMBERSHIP. An unduplicated count of students in membership in a district which is taken, wherever possible, on the date consistent with the special education Child Count in your state (but no earlier than October 1, 2000, and no later than December 15, 2000). The count includes students enrolled in non-district facilities and in pre-kindergarten/pre-school programs.

NON-DISTRICT SCHOOL OR FACILITY. A public or private school or facility that provides instruction or services that are not provided by the local education agency. This includes regional service agencies that provide administrative or special services to local education agency students. A private school may serve children with disabilities who are placed by a public agency in the private school, and who receive special education and related services in the private school at public expense.

CHILDREN WITH DISABILITIES-IDEA. Under the *Individuals with Disabilities Education Act (IDEA),* children with mental retardation, hearing impairments including deafness, speech or language impairments, visual impairments including blindness, emotional disturbance, orthopedic impairments, autism, traumatic brain injury, other health impairments, specific learning disabilities, deaf-blindness, multiple disabilities, or developmental delay, and who, by reason thereof, need special education and related services.

CHILDREN WITH DISABILITIES-504. An elementary or secondary student with a disability who is being provided with related aids and services under *Section 504 of the Rehabilitation Act of 1973*, as amended, and **is not** being provided with services under the *Individuals with Disabilities Education Act* (*IDEA*).

PREGNANT STUDENT: A childbearing woman who is of school age and either is or was enrolled in school at some time during the previous school year.

SPECIFIC INSTRUCTIONS

Item 1. Number Of Public Schools. This number should be equal to the number of Individual School Report ED-102 form(s) submitted with the District Summary ED-101 for this district.

Item 2. Public School Membership. An unduplicated count of students in membership in the district as of October 1, 2000, or the nearest convenient date prior to December 15, 2000. Whenever possible, report public school membership on the date which is as consistent as possible with your state's special education Child Count.

Item 4. Number Of Children With Disabilities-IDEA. Count only the students in this district who are eligible under the *Individuals with Disabilities Education Act*. This number may not necessarily be the same as the aggregate of students reported on the ED-102 form(s) under *Item 10, (Tables 10.1, 10.2, and 10.3), Children with Disabilities,* and *Item 11, Additional Categories of Children with Disabilities.* Include also students in non-district facilities and pre-kindergarten/pre-school children. Intermediate units are considered non-district facilities. Note: Students in non-district facilities and pre-kindergarten are counted on this form but *not* on the ED-102 form.

Item 8. Testing. Please mark the appropriate bubble regarding district- or state-required tests that students in your district were required to pass in the previous 1999-2000 school year.

SPECIAL NOTE

In Item 5 of the ED-102 form, information is collected regarding charter schools. The instruction for charter schools in Item 5 of the ED-102 form says:

> A charter school is a school providing free public elementary or secondary education to eligible students under a specific charter granted by the state legislature or other appropriate authority and designated by such authority to be a charter school.

Only provide data for charter schools for which you have received a pre-printed form. **Darken** the *YES* bubble if the school is a charter school; otherwise, **darken** the *NO* bubble.

Page 2: ED101 - Instruction Sheet

**U.S. DEPARTMENT OF EDUCATION, OFFICE FOR CIVIL RIGHTS
2000 ELEMENTARY AND SECONDARY SCHOOL
CIVIL RIGHTS COMPLIANCE REPORT
DISTRICT SUMMARY: ED101**

17215   Due Date: February 23, 2001

PLEASE CORRECT OR SUPPLEMENT THE INFORMATION ON THE LABEL IF IT IS INCORRECT OR INCOMPLETE:

District Name:

District Address:
Street or P.O. Box

City/Post Office:

County: State: Zip:

1. Report the total number of public schools in this district:

2. Report public school membership of this district (include students served in out-of-district facilities):

3. How many children are awaiting initial evaluation for special education programs and related services?

4. Report the number of children with disabilities-IDEA. Count only students eligible under the *Individuals With Disabilities Education Act.* Do not count students with disabilities who are solely being served under *Section 504 of the Rehabilitation Act of 1973.*

   Of this number, how many are in each of the following (a-c):
   a. How many children are receiving special education and related services in this district's schools or facilities?

   b. How many children are receiving special education and related services in a *non-district* school or facility?

   Of this number, how many are in each of the following [(1) - (6)]:

   (1) Public residential placement?....................

   (2) Private residential placement?..................

   (3) Private separate school?.........................

   (4) Public elementary or secondary school located in another district?.......................

   (5) Regional service agency or intermediate unit?..

   (6) Homebound/hospital?.............................

   c. How many children have been evaluated as requiring special education and related services, but are not receiving these services?..

5. Among the children reported under 4a. and 4b. above, how many are pre-kindergarten/pre-school children?............

6. Report the number of students who were identified as pregnant, for part or all of the school year, in School Year 1999-2000.

   a. Of this number, how many are not in school this year? (Do not count students who graduated.)................................

7. How many children have been identified as having a disability and are receiving related aids and services solely under *Section 504 of the Rehabilitation Act of 1973?* Do not count children who have been identified as having a disability who are receiving services under the *Individuals with Disabilities Education Act.*..........................

8. Were students in your district required in the previous school year (1999-2000), to pass a district-required or state-required test:

   a. to be promoted from any one grade to the next successive grade?   ○ Yes, required by the district   ○ Yes, required by the state   ○ No

   b. to graduate from high school?   ○ Yes, required by the district   ○ Yes, required by the state   ○ No

CERTIFICATION: *I certify that the information is true and correct to the best of my knowledge and belief. (A willfully false statement is punishable by law [U.S. Code, Title 18, Section 1001]).*

Signature of Superintendent or Authorized Representative

Title

Telephone ( ) -

Fax ( ) -

Date / /

OMB#1870-0500
Expiration Date: 12/31/2001

2000 ELEMENTARY AND SECONDARY SCHOOL
CIVIL RIGHTS COMPLIANCE REPORT
INDIVIDUAL SCHOOL REPORT: ED102

U.S. Department of Education, Office for Civil Rights
Washington, D.C. 20202-1172
Due Date: February 23, 2001

REPORTING REQUIREMENT

This Compliance Report is required by the U.S. Department of Education under Title VI of the Civil Rights Act of 1964, Title IX of the Education Amendments of 1972, and Section 504 of the Rehabilitation Act of 1973. Implementing regulations are issued to carry out the purposes of Title VI of the Civil Rights Act of 1964, at 34 CFR 100.6(b); Title IX regulations at 34 CFR *106.71*; and the Section 504 regulations are at 34 CFR 104.61.

Public Burden Statement. According to the Paperwork Reduction Act of 1995, no persons are required to respond to a collection of information unless such collection displays a valid OMB control number. The valid OMB control number for this information collection is 1870-0500. The time required to complete this information collection is estimated to average 9 hours per response, including the time to review instructions, research existing data resources, gather the data needed, and complete and review the information collection. **If you have any comments concerning the accuracy of the time estimate(s) or suggestions for improving this form, please write to:** U.S. Department of Education, Washington, D.C. 20202-1172. **If you have comments or concerns regarding the status of your individual submission of this form, write directly to:** Compliance Reports Coordinator, U.S. Department of Education, 400 Maryland Avenue, S.W., Room 5316, Switzer Building, Washington, D.C. 20202-1172.

GENERAL INSTRUCTIONS

- This form should be completed for each public school in the district.

- Please print legibly using a black ball-point pen.

- If you select to report via a mainframe computer cartridge or PC diskette, please see separate documentation.

- Information should be reported as of October 1, 2000, or the nearest convenient date prior to December 15, 2000, unless otherwise noted. Whenever possible, information should be provided consistent with the date of the special education Child Count in your state.

- **In order to allow us to distinguish between no students for a given item and non-applicability of that item for your school, please enter a response to all items.** If the answer to a given item is *NONE*, enter *zero (0)* in the appropriate space. If a particular item is not applicable in your case, enter an X in the "total" column if the item has a total; otherwise, enter an *X* in the far right column for that item.

- **Please mail original forms to the Office for Civil Rights.** Retain photocopied ED-101 and ED-102 forms for your reference for two (2) years from the date signed. The mailing address is: Compliance Report Project Office, Office for Civil Rights, 400 Maryland Avenue, S.W., Washington, D.C. 20202-1172.

- The "Optional TOTAL" in Column 6 of Tables 10.1, 10.2, and 10.3, and Column 4 and Row J of Table 11 are not required by OCR. They are intended for your use in making calculations if you choose to do so.

- Please add the telephone number and the **FAX telephone number** in the appropriate boxes on the form for the individual in your school who can respond to questions regarding this form.

- The certification signature block located on page 5 must be completed for all schools. **It is essential that all ten (10) pages be returned, even if no items are completed on pages 6, 7, 8, or 9.** The certification pertains to all items on the ED-102 form.

- **All schools must complete Item 17 (Teachers) on page 10 of the form.**

Page 1: ED102 - Instruction Sheet

DEFINITIONS

PUBLIC SCHOOL. An institution that provides pre-school, elementary and/or secondary instruction; has one or more grade groupings *(pre-kindergarten through 12)* or is ungraded; has one or more teachers to give instruction; is located in one or more buildings; has an assigned administrator(s); receives public funds as its primary support; and is operated by an education agency. Public schools include charter schools that receive public funding from local or state sources.

PUBLIC SCHOOL ENROLLMENT. An unduplicated count of students enrolled in the district as of October 1, 2000, or the nearest convenient date prior to December 15, 2000. Whenever possible, report public school enrollment on the date which is as consistent as possible with the special education Child Count date in your state.

CHILDREN WITH DISABILITIES-IDEA. Under the *Individuals with Disabilities Education Act (IDEA),* children with mental retardation, hearing impairments including deafness, speech or language impairments, visual impairments including blindness, emotional disturbance, orthopedic impairments, autism, traumatic brain injury, other health impairments, or specific learning disabilities, deaf-blindness, multiple disabilities, or developmental delay; and who, by reason thereof, need special education and related services.

CHILDREN WITH DISABILITIES-504. An elementary or secondary student with a disability who is being provided with related aids and services under *Section 504 of the Rehabilitation Act of 1973*, as amended, and **is not** being provided with services under the *Individuals with Disabilities Education Act (IDEA)*.

■ Children receiving special education services under the *Individuals with Disabilities Education Act (IDEA),* defined under Children with Disabilities-IDEA above, are reported in the column *"Served under IDEA"* in Table 9, or *"Students with Disabilities/IDEA"* in Tables 12A and 12B, or *"Students with Disabilities-IDEA"'* in Tables 7, 14, and 15. Children receiving services under *Section 504 of the Rehabilitation Act of 1973,* as amended, are reported in the column *"Served under Section 504 Only"* in Table 9 and in the column *"Section 504 Only"* in Tables 12A and 12B.

ABILITY GROUPING. Pedagogical practice of separating students into different classrooms within a grade based on their estimated achievement or ability levels, and who are ability grouped for classroom instruction in mathematics, or English-Reading-Language Arts.

> **NOTE ONE:** In this application, ability grouping does NOT include grouping by achievement level on the basis of required prerequisites for certain courses, i.e., Algebra I as a prerequisite for Algebra II.

> **NOTE TWO:** Ability grouping includes students pulled out of their regular mathematics, or English-Reading-Language Arts classes for Title I purposes in these subject areas.

RACE/ETHNICITY CATEGORIES

■ *American Indian or Alaskan Native:* A person having origins in any of the original peoples of North America and who maintains cultural identification through tribal affiliation or community recognition.

■ *Asian or Pacific Islander:* A person having origins in any of the original peoples of the Far East, Southeast Asia, the Pacific Islands, or the Indian subcontinent. This includes, for example, China, India, Japan, Korea, the Philippine Islands, and Samoa.

■ *Hispanic:* A person of Mexican, Puerto Rican, Cuban, Central or South American, or other Spanish culture or origin regardless of race.

■ *Black (Not of Hispanic Origin):* A person having origins in any of the Black racial groups of Africa.

■ *White (Not of Hispanic Origin):* A person having origins in any of the original peoples of Europe, North Africa, or the Middle East.

> **NOTE:** In October 1997, the Office of Management and Budget (OMB) announced its decision concerning the revision of the standards for Federal data on race and ethnicity. In that announcement, OMB reported that there would be five racial categories -- American Indian or Alaska Native, Black or African American, Asian, Native Hawaiian or Other Pacific Islander, and White -- and one ethnic category -- Hispanic or Latino. Additionally, OMB announced that individuals would be allowed to select one or more categories. Under the new reporting requirements, a single, multi-racial category can **not** be used. OCR is currently working with OMB and other program offices in the U.S. Department of Education to develop reporting categories for aggregating multiple race responses. OCR expects to use these categories in the coming years in future versions of this Compliance Report. The Office for Civil Rights will provide ample notice to public elementary and secondary schools before these revisions go into effect.

Page 2: ED102 - Instruction Sheet

LIMITED ENGLISH PROFICIENT (LEP) STUDENT. (1) Individuals who were not born in the United States or whose native language is a language other than English; (2) individuals who come from environments where a language other than English is dominant; and (3) individuals who are American Indians and Alaskan Natives and who come from environments where a language other than English has had a significant impact on their level of English language proficiency; and who, by reason thereof, have sufficient difficulty speaking, reading, writing, or understanding the English language, to deny such individuals the opportunity to learn successfully in classrooms where the language of instruction is English or to participate fully in our society.

- The *LEP* column in Tables 7, 8, 10.1, 10.2, 10.3, 12A, 12B, 13, 14, and 15 means the number of students needing LEP programs.

> **NOTE:** **The three definitions which follow (which are used in Tables 10.1, 10.2, 10.3, and 11) are consistent with definitions used by the Office of Special Education Programs Placement form.**

CHILDREN WHO RECEIVED SPECIAL EDUCATION OUTSIDE THE REGULAR CLASS LESS THAN 21 PERCENT OF THE SCHOOL DAY. The number of children with disabilities receiving special education and related services outside the regular classroom for less than 21 percent of the school day. This may include children with disabilities placed in: regular class with special education/related services provided within regular classes; regular class with special education/related services provided outside regular classes; or regular class with special education services provided in resource rooms.

CHILDREN WHO RECEIVED SPECIAL EDUCATION OUTSIDE THE REGULAR CLASS AT LEAST 21 PERCENT BUT NO MORE THAN 60 PERCENT OF THE SCHOOL DAY. The number of children with disabilities receiving special education and related services outside the regular classroom for at least 21 percent but no more than 60 percent of the school day. This may include: resource rooms with special education/related services provided within the resource room; or resource rooms with part-time instruction in a regular class.

CHILDREN WHO RECEIVED SPECIAL EDUCATION OUTSIDE REGULAR CLASS FOR MORE THAN 60 PERCENT OF THE SCHOOL DAY. The number of children with disabilities receiving special education and related services outside the regular classroom for more than 60 percent of the school day. Do not include children who receive education programs in separate day or residential facilities. This category may include children placed in: self-contained classrooms with part-time instruction in a regular class or self-contained special classrooms with full-time special education instruction on a regular school campus.

SPECIFIC INSTRUCTIONS

Item 1. Grades Offered. Darken the appropriate *YES* or *NO* bubble for each grade offered in this school. Also **darken** the bubble which represents the level that you consider your school to be. If you consider your school to be other than an elementary, middle/junior, or high school (for example, a school which offers instruction at more than one of these levels), please **darken** the *OTHER* bubble. If your school is totally ungraded, **darken** the *totally ungraded* bubble.

Item 2. Special Education. Darken the *YES* bubble if this school offers *only* special education classes, otherwise **darken** the *NO* bubble.

Item 3. Ability Grouping. Darken the *YES* bubble if you have any students in this school who are ability grouped for classroom instruction in mathematics or English-Reading-Language Arts; otherwise, **darken** the *NO* bubble.

> **NOTE ONE:** In this application, ability grouping does NOT include grouping by achievement level on the basis of required prerequisites for certain courses, i.e., Algebra I as a prerequisite for Algebra II.

> **NOTE TWO:** Ability grouping includes students pulled out of their regular mathematics, or English-Reading-Language Arts classes for Title I purposes in these subject areas.

Item 4. Magnet School or Program. A magnet school or program is a special school or program designed to attract students of different racial/ethnic backgrounds for the purpose of reducing, preventing or eliminating racial isolation. Racial isolation means a school with 50 percent or more minority enrollment. **Darken** the appropriate bubble, if this school is a magnet school or has a magnet program, regardless of the source of funding, i.e., Federal, state, or local government.

Item 5. Charter School. A charter school is a school providing free public elementary or secondary education to eligible students under a specific charter granted by the state legislature or other appropriate authority and designated by such authority to be a charter school. **Only provide data for charter schools for which you have received a pre-printed form. Darken** the *YES* bubble if the school is a charter school; otherwise, **darken** the *NO* bubble.

Page 3: ED102 - Instruction Sheet

Item 6. Alternative School. An alternative school is a public elementary or secondary school that addresses the needs of students which typically cannot be met in a regular school and provides nontraditional education which falls outside of the categories of regular education, special education, vocational education, gifted and talented or magnet school programs. This definition includes schools which are adjunct to a regular school, e.g., are located on the same campus as a regular school but have a separate principal or administrator. **Darken** the *YES* bubble if this school is an alternative school; otherwise, **darken** the *NO* bubble. Also **darken** as many bubbles as are appropriate if the school is designed to meet the needs of pregnant students, students with academic difficulties, and/or students with discipline problems.

Item 7. Pupil Statistics. (Do not include *pre-kindergarten/pre-school* children).

> **NOTE:** The column *"Students with Disabilities: IDEA"* in this table means children with disabilities receiving special education services under the *Individuals with Disabilities Education Act*. The column *"LEP"* in this table means the number of students needing LEP programs.

A. Enrollment. Enter in Table 7, Row A the unduplicated count of students on the rolls of the school taken, whenever possible, as of the date which is consistent with the date of the special education Child Count in your state (but no earlier than October 1, 2000, and no later than December 15, 2000). **The total number of male and female students in the *Students with Disabilities:IDEA* column (column 7) should be equal to the sum of the totals reported in Tables 10.1, 10.2, 10.3, and 11.**

B. In Gifted Or Talented (G/T) Programs. Enter in Table 7, Row B the number of students enrolled in special programs during regular school hours for students who possess unusually high academic ability or a specialized talent or aptitude such as in literature or the arts. Count students once regardless of the number of classes in which they are enrolled.

C. Needing LEP Programs. Enter in Table 7, Row C the number of students who have a home language other than English and who are so limited in their English proficiency that they cannot participate meaningfully in the school's regular instructional program.

D. Enrolled In LEP Programs. Enter in Table 7, Row D the number of students reported in Table 7, Row C as needing LEP programs who are enrolled in a program of language assistance (e.g., English-as-a-Second-Language or bilingual education). Do not count students enrolled in a class to learn a language other than English.

Item 8. Discipline of Students without Disabilities.

> **NOTE:** Discipline of Students with Disabilities is reported in Table 9, using definitions of long-term suspension which correspond to those used in the *Individuals with Disabilities Education Act*. Data for students without disabilities should use the following definitions.

A. Corporal Punishment. Enter in Table 8, Row A the number of students who received corporal punishment during the previous (1999-2000) school year. *Corporal punishment* is paddling, spanking, or other forms of physical punishment imposed on a student. If your state or school district has a policy banning corporal punishment, please enter X in the far right box to indicate that this item does not apply. Count each student only once regardless of the number of times he or she was punished.

B. Out-of-School Suspensions. Enter in Table 8, Row B the number of students suspended from school for at least one (1) day during the previous (1999-2000) school year. *Out-of-School Suspension* is defined as excluding a student from school for disciplinary reasons for one school day or longer. Count students only once regardless of the number of times suspended. Do not count students suspended from the classroom who served the suspension in the school.

C. Total Expulsions. Enter in Table 8 Row C, the number of students expelled from school during the previous (1999-2000) school year. An *expulsion* is defined as the exclusion from school for disciplinary reasons that results in the student's removal from school attendance rolls or that meets the criteria for expulsion as defined by the appropriate state or local school authority. This includes expulsions where the student, although expelled from a particular school, continues to receive educational services from the district. Do *not* include suspensions.

D. Expulsions--Total Cessation of Educational Services. Of the total number of students expelled from school during the previous (1999-2000) school year (Table 8 Row C), enter the number who had a total cessation of educational services--that is, the student, after expulsion from a school, was not offered other educational services by either the school or the district.

Page 4 - ED102 Instruction Sheet

E. **Expulsions--Zero Tolerance Policies.** Of the total number of students expelled during the previous (1999-2000) school year (Table 8 Row C), enter the number of students who were expelled because of zero tolerance policies. A zero tolerance expulsion policy is a policy that results in mandatory expulsion of any student who commits one or more specified offenses (for example, some zero tolerance policies specify offenses involving guns, or other weapons, or violence, or similar factors, or combinations of these factors).

> **NOTE:** A zero tolerance expulsion policy should still be included in your response to this question, even if the resulting "mandatory" expulsion may be subject to some narrow or limited exceptions. For example, the Federal Gun -Free Schools Act permits "State law to allow the chief administering officer of . . . a local education agency to modify such expulsion requirement for a student on a case-by-case basis", and State or district zero tolerance expulsion policies may similarly give discretion for limited exceptions to the strict application of the expulsion requirement. Such policies would still be "zero tolerance policies" which should be included in your responses to this question. The count requested should only include students actually expelled as a result of such policies.

Item 9. Discipline of Students with Disabilities. Schools must report data on the numbers of students receiving corporal punishment during the 1999-2000 school year in Row A (see Item 8 for definition of corporal punishment). Schools also must report data on the number of students with disabilities who received long-term suspensions/expulsions during the 1999-2000 school year (suspensions of more than 10 days) in Row B (students who continued to receive services) and Row C (students for whom there was a cessation of services). The column entitled "*Served under Section 504 Only*" refers to those students receiving services solely under *Section 504 of the Rehabilitation Act of 1973, as amended.* In each row, report students only once regardless of the number of times he or she was disciplined. See general instructions for the appropriate response if a cell has no students or is not applicable to this school. Individual students may be reported in more than one row.

Item 10. Children with Disabilities. (Do *not* count *pre-kindergarten/pre-school* children.) Enter the number of students with disabilities by race/ethnicity, educational placement, sex, and LEP. Report all students receiving special education services at this school, whether or not they reside in this school district. Count each student only once. If a student has more than one disability, count by the primary disability. Please read the instructions fully before completing this item.

Enter in Table 10.1:

MENTAL RETARDATION. This refers to significantly sub-average general intellectual functioning existing concurrently with deficits in adaptive behavior and manifested during the developmental period, which adversely affect a child's educational performance.

A. *Mild Retardation.* Students require intermittent support to perform functional academic skills, activities of daily living (self-care, home living, use of their community, recreation and leisure activities, work) or communicating and interacting with others. This support may be episodic, time-limited (may be intense but for a relatively short period of time), or of low intensity over a long period of time.

B. *Moderate Retardation.* Students require limited but continuing support to perform functional academic skills, activities of daily living (self-care, home living, use of their community, recreation and leisure activities, work) or communicating and interacting with others. This support may be consistent over time. It may be either time-limited (but may be intense for a substantial period of time), or of low intensity over a life span.

C. Severe *Retardation.* Students require extensive or pervasive support to perform functional academic skills, activities of daily living (self-care, home living, use of their community, recreation and leisure activities, work) or communicating and interacting with others. Support may be of high intensity, over long periods of time, or potentially life sustaining.

> **NOTE:** Complete Rows A, B, and C if your school collects this information. Rows A, B, and C are optional if your school does **not** already collect this information. If a particular cell is not applicable, enter an X in the right-most box of the cell.

D. *Total is the total of Table 10.1 Rows A, B, and C.* You must complete Row D, regardless of whether or not you have reported data in Rows A, B, and C. If a particular cell is not applicable in your case, enter an X in the right-most box of the cell.

> **NOTE:** The computational total in column 6 of this table is optional. You may complete it if it will assist you in your computations. It is not required by OCR.

Page 5: ED102 - Instruction Sheet

Enter in Table 10.2:

> **NOTE:** The definitions of disability categories which follow are the same as the definitions used by the Office of Special Education Programs, as specified in the regulations for the *Individuals with Disabilities Education Act.*

A. *Emotional Disturbance.* [previously entitled Serious Emotional Disturbance] This refers to a condition exhibiting one or more of the following characteristics over a long period of time and to a marked degree, which adversely affects a child's educational performance: (1) an inability to learn, which cannot be explained by intellectual, sensory, or health factors; (2) an inability to build or maintain satisfactory interpersonal relationships with peers and teachers; (3) inappropriate behavior or feelings under normal circumstances; (4) a general pervasive mood of unhappiness or depression; or (5) a tendency to develop physical symptoms or fears associated with personal or school problems. The term includes schizophrenia. The term does not apply to children who are socially maladjusted, unless it is determined that they have an emotional disturbance.

B. *Specific Learning Disability.* This refers to a disorder in one or more of the basic psychological processes involved in understanding or in using language, spoken or written, which may manifest itself in an imperfect ability to listen, think, speak, read, write, spell, or do mathematical calculations. The term includes such conditions as perceptual disabilities, brain injury, minimal brain dysfunction, dyslexia, and developmental aphasia. The term does not include learning problems that are primarily the result of visual, hearing, or motor disabilities, of mental retardation, of emotional disturbance, or of environmental, cultural, or economic disadvantage.

> **NOTE:** The computational total in column 6 is optional. You may complete it if it will assist you in your computations. It is not required by OCR.

Enter in Table 10.3: Developmental Delay.

Developmental Delay as defined in the *Individuals with Disabilities Education Act* is defined as a child who is experiencing developmental delays, as defined by your state, and as measured by appropriate diagnostic instruments and procedures in one or more of the following cognitive areas: physical development, cognitive development, communication development, social or emotional development, or adaptive development. Please refer to the instructions on the table for this item before you complete it, in order to ensure that your state and your district meet all necessary requirements.

> **NOTE:** The computational total in column 6 is optional. You may complete it if it will assist you in your computations. It is not required by OCR.

> **NOTE:** The column *LEP* in these tables means the number of students needing LEP programs.

Item 11. Additional Categories of Children with Disabilities. (Do *not* include *pre-kindergarten/pre-school* children). Enter the number of students by educational placement and by disability. Report all students receiving special education services at this school, whether or not they reside in this school district. Count each student only once. If a student has more than one disability, count by the primary disability.

> **NOTE:** The computational total in column 4 is optional. You may complete it if it will assist you in your computations. It is not required by OCR.

A. *Hearing Impairments.* This refers to an impairment in hearing, whether permanent or fluctuating, that adversely affects a child's educational performance. It also includes a hearing impairment that is so severe that the child is impaired in processing linguistic information through hearing, with or without amplification, that adversely affects a child's educational performance.

B. *Speech or Language Impairments.* This refers to a communication disorder, such as stuttering, impaired articulation, a language impairment, or a voice impairment, that adversely affects a child's educational performance.

C. *Visual Impairments.* This refers to a visual impairment which, even with correction, adversely impacts a child's educational performance. The term includes both partial sight and blindness.

D. *Orthopedic Impairments.* This refers to a severe orthopedic impairment that adversely affects a child's educational performance. The term includes impairments caused by congenital anomaly (e.g., clubfoot, absence of some member, etc.), impairments caused by disease (e.g,. poliomyelitis, bone tuberculosis, etc.) and impairments from other causes (e.g., cerebral palsy, amputations, and fractures or burns that cause contractures).

Page 6: ED102 - Instruction Sheet

E. *Autism.* This refers to a development disability significantly affecting verbal and non-verbal communication and social interaction, generally evident before age 3, that adversely affects educational performance. Other characteristics often associated with autism are engagement in repetitive activities and stereotyped movements, resistance to environmental change or change in daily routines, and unusual responses to sensory experiences. Autism doesn't apply if a child's educational performance is adversely affected primarily because the child has an emotional disturbance.

F. *Traumatic Brain Injury.* This refers to an acquired injury to the brain caused by an external physical force, resulting in total or partial functional disability or psychosocial impairment or both, that adversely affects a child's educational performance. The term applies to open or closed head injuries resulting in impairments in one or more areas, such as cognition; language; memory; attention; reasoning; abstract thinking; judgement; problem-solving; sensory, perceptual, and motor abilities; psychosocial behavior; physical functions; information processing; and speech. The term does not apply to brain injuries that are congenital or degenerative, or brain injuries induced by birth trauma.

G. *Deaf-Blindness.* This refers to concomitant hearing and visual impairments, the combination of which causes such severe communication and other developmental and educational problems that they cannot be accommodated in special education programs solely for children with blindness or children with deafness.

H. *Multiple Disabilities.* This refers to concomitant impairments (such as mental retardation-blindness, mental retardation-orthopedic impairments, etc.), the combination of which causes such severe educational problems that the problems cannot be accommodated in special education programs solely for one of the impairments. The term does not include deaf-blindness.

I. *Other Health Impairments.* This refers to having limited strength, vitality, or alertness, due to chronic or acute health problems such as a heart condition, tuberculosis, rheumatic fever, nephritis, asthma, sickle cell anemia, hemophilia, epilepsy, lead poisoning, leukemia or diabetes, which adversely affects a child's educational performance.

J. *Total.* This is an optional computational row. You may complete it if it will help you in your computation. It is not required by OCR.

Note on maintaining data by sex on students with disabilities: The sum of the totals reported in Tables 10.1, 10.2, 10.3 and 11 equals the number of Students With Disabilities receiving special education services under IDEA reported in Column 7 on Table 7, Row A. A student should be counted only once, based on the child's primary disability, and in only one of the four tables. Although you are not required to report data by sex on Table 11, you are required to maintain data on the sex of all students with disabilities for the purposes of responding to Table 7 Row A. Please note that districts are only required to provide data on the sex of specific subcategories of students with disabilities in Tables 10.1, 10.2, and 10.3.

SPECIFIC INSTRUCTIONS FOR TABLES 12A AND 12B: TESTING

TABLE 12A IS TO BE COMPLETED BY ELEMENTARY AND MIDDLE SCHOOLS (GRADES K-8)

Table 12A. Testing (for Grade-to-grade promotion). Please complete this table *if your school administered, in the 1999-2000 school year, a district- or state-required test that students are either required to pass or that is used as a significant factor in making promotion decisions for all students taking the test.* If your school conducted tests for grade-to-grade promotion for more than one grade, please photocopy the table as many times as are necessary BEFORE filling it out in order to report on each test. Report data using both the original table on this page and as many photocopied tables as are appropriate.

If students were **not** required to pass a district- or state-required test to be promoted from one grade to the next, please **darken** the bubble entitled *"No such tests were administered".*

If students were required to pass such a test, please **darken** the appropriate bubble indicating whether this test was a "sole criterion" or a "significant criterion", and complete the table. If all students were required to take a district- or state-required test, and must pass the test to be promoted from one grade to the next, please **darken** the bubble entitled *"Sole criterion".* However, if all students were required to take the test, and the test is an important criterion in the decision on whether or not to promote the student from grade to grade, but other criteria, such as teacher recommendations or student grades were used in the promotion decision, please **darken** the bubble entitled *"Significant criterion".*

Please provide the following data for the testing of students in these grades during the 1999-2000 school year, by race/ethnicity, limited English proficiency (in the column entitled *LEP),* and whether the student is receiving services under the *Individuals with Disabilities Education Act* (in the column entitled *Students with Disabilities-IDEA)* or under *Section 504 of the Rehabilitation Act of 1973* (in the column entitled *Section 504 Only),* and sex.

Page 7: ED102 - Instruction Sheet

Do not count students who were not tested because they passed the test on a previous occasion.

Include in Rows A or B those students who took the test and were provided with accommodations, modifications, or adaptations, such as a different setting, extended time, Braille, or use of dictionaries by LEP students.

All students who were excluded from taking a test for grade-to-grade promotion and who did not take an alternate assessment should be reported in Row C.

Students who were tested using alternate assessments should be reported in Row D. An alternate assessment is an assessment provided to children with disabilities who cannot participate in a state- or district-wide assessment program, even with appropriate accommodations.

If students are required to pass more than one test in order to be promoted from one grade to the next, include that student in the row entitled *Tested and Passed* if that student passed all tests that he or she was required to pass; if the student failed one or more tests, report that student in the row entitled *Tested and Failed.*

TABLE 12B IS TO BE COMPLETED BY HIGH SCHOOLS (GRADES 9-12)

Table 12B. Testing (for Graduation from high school). Please complete this table *if your school administered, in the 1999-2000 school year, a district- or state-required test that students are either required to pass or that is used as a significant factor in making graduation decisions for all students taking the test.*

If students were **not** required to pass a district- or state-required test to graduate from high school, please **darken** the bubble entitled *"No such tests were administered".* If students were required to pass such a test, please **darken** the appropriate bubble indicating whether this test was the "sole criterion" or a "significant criterion", and complete the table.
If all students were required to take a district- or state-required test, and must pass the test to graduate from high school, please **darken** the bubble entitled *"Sole criterion".* However, if all students were required to take the test, and the test is an important criterion in the decision on whether or not the student graduates from high school, but such other criteria as teacher recommendations or student grades were used in the graduation decision, please **darken** the bubble entitled *"Significant criterion".*

Please provide the following data for the testing of students during the 1999-2000 school year, by race/ethnicity, limited English proficiency (in the column entitled *LEP),* and whether the student is receiving services under the *Individuals with Disabilities Education Act* (in the column entitled *Students with Disabilities-IDEA)* or under *Section 504 of the Rehabilitation Act of 1973* (in the column entitled *Section 504 Only),* and sex.

Do not count students who were not tested because they passed the test on a previous occasion.

In Rows A or B, include those students who took the test and were provided with accommodations, modifications, or adaptations, such as a different setting, extended time, Braille, or use of dictionaries by LEP students.

All students who were excluded from taking a test for graduation from high school and who did not take an alternate assessment should be reported in Row C.

Students who were tested using alternate assessments should be reported in Row D. An alternate assessment is an assessment provided to children with disabilities who cannot participate in a state- or district-wide assessment program, even with appropriate accommodations.

If students are required to pass more than one test in order to graduate from high school, include that student in the row entitled *Tested and Passed* if that student passed all tests that he or she was required to pass; if the student failed one or more tests, report that student in the row entitled *Tested and Failed.*

ITEM 13 IS TO BE COMPLETED FOR THE HIGHEST AND LOWEST ELEMENTARY GRADES
(BETWEEN GRADES 1 AND 6) ONLY

Item 13. Student Assignment. Complete this table only if the total percentage of minority students (American Indian or Alaskan Native, Asian or Pacific Islander, Hispanic, and Black [Not of Hispanic Origin]) in this school is more than 20 percent but less than 80 percent. (Do not include *pre-kindergarten/pre-school* or *kindergarten*). Report only the **entry (lowest) or exit (highest) elementary grades, which are typically grades one and five or six**. Enter in Table 13 the grade level and **darken** the bubble under the respective *YES* or *NO* columns if students are grouped in that class according to ability level. For the ability grouping definition to be used in completing Table 13, please refer to Item 3 of the Specific Instructions. Please complete by race/ethnicity and limited English proficiency.

Page 8: ED102 - Instruction Sheet

Item 14. Advanced Placement. Enter the number of students by race/ethnicity, sex, LEP, and disability status (students receiving services under the *Individuals with Disabilities Education Act*) who are currently enrolled in Advanced Placement Program mathematics or science courses. Mathematics includes calculus AB and BC. Science includes biology, chemistry, and physics. If this school does *not* participate in a particular Advanced Placement Program course, **darken** the bubble in the *Not Offered* column for that course. If the school does not offer any Advanced Placement Programs for high school students please **darken** the *Not Offered* bubble for Table 14. Type of AP class means the particular AP course, i.e., mathematics, English, computer science, etc, --and *not* the number of AP classes offered.

Item 15. High School Completers. Enter the number of students who received a regular high school diploma or a certificate of attendance or completion from the previous (1999-2000) school year. Certificate of attendance or completion refers to an award of less than a regular diploma, or a modified diploma, or fulfillment of an Individual Education Plan for students with disabilities. Please complete by race/ethnicity, sex, LEP, and disability status (students receiving services under the *Individuals with Disabilities Education Act*).

Item 16. Interscholastic Athletics. For the entire previous school year (1999-2000), enter the number of sports, teams, and students as of the day of the *first* official interscholastic competition (e.g., game, match, meet). Do *not* include intramural sports or cheerleading. Count each competitive level of a given sport as a separate item (e.g., freshman, junior varsity, and varsity). For example, basketball is *one* sport, but there may be more than one basketball team (e.g., varsity boys, varsity girls, junior varsity boys, etc.). Count a student once for each team he or she is on. For example, a student should be counted *twice* if he or she is on *two* teams.

ITEM 17 IS TO BE COMPLETED BY ALL SCHOOLS

Item 17. Teachers. For item 17(a), enter the total number of full-time teachers employed by your school on October 1, 2000. For item 17(b), enter the number of full-time teachers employed by your school who meet all applicable state teacher certification requirements for a standard certificate. Do not include teachers who have emergency, temporary or provisional certification. For beginning teachers, include, as certified, those who have met all of the standard teacher education requirements with the exception of the State-required probationary period.

You must return all sheets of this form, even if you did not use pages 6, 7, 8, or 9 because it did not pertain to your school.

Page 9: ED102 - Instruction Sheet

U.S. DEPARTMENT OF EDUCATION, OFFICE FOR CIVIL RIGHTS
2000 ELEMENTARY AND SECONDARY SCHOOL
CIVIL RIGHTS COMPLIANCE REPORT
INDIVIDUAL SCHOOL REPORT: ED102   DUE DATE: February 23, 2001

OMB No. 1870-0500
Expiration Date: 12/31/2001

*17300

PLEASE CORRECT OR
SUPPLEMENT THE
INFORMATION ON THE
LABEL IF IT IS INCORRECT
OR INCOMPLETE:

District Name:

School Name:

School Address:
Street or P.O. Box

City/Post Office:

State: Zip:

County:

1. **Grades Offered:** (Darken all that apply.)
   If this school is totally ungraded, darken this bubble: ◉

   If this school is partially or totally graded, darken the grades offered in the bubbles below:

	Pre-K	K	1	2	3	4	5	6	7	8	9	10	11	12
YES	○	○	○	○	○	○	○	○	○	○	○	○	○	○
NO	○	○	○	○	○	○	○	○	○	○	○	○	○	○

   What is the level of this school?
   ○ Elementary   ○ High School
   ○ Middle/Junior High ○ Other

2. **Special Education:** Does this school offer **only** special education services?........................ ○ Yes ○ No

3. **Ability Grouping:** Do you have any students in this school who are ability grouped for classroom instruction in mathematics or English-Reading-Language Arts?.... ○ Yes ○ No

4. **Magnet School or Program:**
   a. Is this school either a magnet school or a school operating a magnet program within the school?................. ○ Yes ○ No
   b. **If the answer to 4a was "YES"**, does the entire school population participate in the magnet school program?................. ○ Yes ○ No

5. **Charter School:** Is this school a charter school?.......... ○ Yes ○ No If so, is it designed to meet the needs of: ○ students with academic difficulties? ○ students with discipline problems?
   ○ pregnant students?

6. **Alternative School:** Is this school an alternative school? ○ Yes ○ No

7. **Pupil Statistics:** (Do not include pre-kindergarten/pre-school children.) The column *Students with Disabilities: IDEA* refers to children and youth receiving services under the *Individuals with Disabilities Education Act*.

* SEX:
M = MALE;
F = FEMALE
**LEP = limited English proficient

TABLE 7 Pupil Statistics

NUMBER OF STUDENTS	SEX*	RACE/ETHNICITY					(6) TOTAL	(7) Students with Disabilities: IDEA	(8) LEP**
		(1) American Indian or Alaskan Native	(2) Asian or Pacific Islander	(3) Hispanic	(4) Black, Not of Hispanic Origin	(5) White, Not of Hispanic Origin			
A. Enrollment	M								
	F								
B. In Gifted/ Talented Programs	M								
	F								
C. Needing LEP Programs	M								
	F								
D. Enrolled in LEP Programs (Of those reported in 7.C.)	M								
	F								

The sum of the totals reported in Tables 10.1, 10.2, 10.3 and 11 equals the number of Students With Disabilities receiving special education services under IDEA reported in Column 7 on Table 7. Row A. Although you are not required to report data by sex on Table 11, you are required to maintain data on the sex of all students with disabilities for the purposes of responding to Table 7 Row A. Please note that districts are only required to provide data on the sex of specific subcategories of students with disabilities in Tables 10.1, 10.2, and 10.3.

ORIGINAL - Return to Office for Civil Rights (LEGAL COPY)

District Name: _____

School Name: _____

Individual School Report: ED102 — Page 2 of 10

Table 8: Discipline (Students Without Disabilities). Please report the following data for the 1999 - 2000 School Year. Do not include pre-kindergarten/pre-school children. **Please read the specific instructions for this table carefully before completing this item.**

TABLE 8

NUMBER OF STUDENTS	SEX*	RACE/ETHNICITY						
		(1) American Indian or Alaskan Native	(2) Asian or Pacific Islander	(3) Hispanic	(4) Black, Not of Hispanic Origin	(5) White, Not of Hispanic Origin	(6) TOTAL	(7) LEP**
A. Corporal Punishment	M							
	F							
B. Out of School Suspensions	M							
	F							
C. Total Expulsions	M							
	F							
D. Expulsions -- Total Cessation of Educational Services	M							
	F							
E. Expulsions -- Zero Tolerance Policies	M							
	F							

* SEX:
M = MALE;
F = FEMALE
**LEP = limited English proficient

Table 9: Discipline of Students with Disabilities. Please report, for the 1999 - 2000 School Year, data on corporal punishment, and long-term suspensions/expulsions (both with and without services) for students with disabilities served under the *Individuals with Disabilities Education Act* (in the column entitled *Served under IDEA*) and *Section 504 of the Rehabilitation Act of 1973* (in the column entitled *Served under Section 504 Only*). See the specific instructions for the definition of long-term suspension/expulsion and the specific instructions for Item 8 for the definition of corporal punishment. For each row, count each student only once. Do not include pre-kindergarten/pre-school children. See general instructions for the appropriate response if a cell has no students or is not applicable to this school. Individual students may be reported in more than one row.

TABLE 9

NUMBER OF STUDENTS	(1) Served under IDEA	(2) Served under Section 504 Only
A. Corporal Punishment		
B. Long-term suspension/expulsion: non-cessation of services		
C. Long-term suspension/expulsion: cessation of services		

ORIGINAL - Return to Office for Civil Rights (LEGAL COPY)

17300

District Name: _____

School Name: _____

Individual School Report: ED102 — Page 3 of 10

Item 10: Children with Disabilities. Please complete the following tables by race, sex, LEP and educational placement for the number of children with disabilities receiving services who, under the *Individuals with Disabilities Education Act*, are receiving services in this school. Educational placement is defined as the percentage of the day that a student receives special education services outside the regular class. Include all students attending this school regardless of whether they are resident or non-resident of the reporting school district. Do *not* count *pre-kindergarten/pre-school* children. See general instructions for the appropriate response if a cell has no students or is not applicable to this school (e.g., this school does not use the subcategories of mild, moderate, and severe).

TABLE 10.1

DISABILITY CATEGORY (See Specific Instructions, for Item 10, before completing this item.)	SEX*	RACE/ETHNICITY					Optional**	EDUCATIONAL PLACEMENT: TIME OUTSIDE REGULAR CLASSROOM			
		(1) American Indian or Alaskan Native	(2) Asian or Pacific Islander	(3) Hispanic	(4) Black, Not of Hispanic Origin	(5) White, Not of Hispanic Origin	(6) TOTAL	(7) Less than 21%***	(8) Between 21% and 60%***	(9) More than 60%***	(10) LEP****
A. Mild Retardation	M										
	F										
B. Moderate Retardation	M										
	F										
C. Severe Retardation	M										
	F										
D. TOTAL	M										
	F										

* SEX: M = MALE; F = FEMALE

** Dark lines denote that this part of the table is optional. (See page 5 of instruction sheet.)

*** See Definitions on page 3 of ED102 instruction sheet for definitions, including complete label for *Less than 21%*, *Between 21% and 60%, and More than 60%*

**** LEP = limited English proficient

PLEASE FILL IN ALL INFORMATION INCLUDING LEP ITEMS. IF THE ANSWER IS ZERO, PLEASE ENTER ZERO. (For example, if you entered "school total" LEP data other than zero in Table 7 (Table 7, Row C, Column 6), but your school has no LEP students in the more specific "sub-total" categories on this page, please enter zero. If you have LEP students in these subcategories, please enter the number of these students.)

District Name: _____

School Name: _____

Individual School Report: ED102 — Page 4 of 10

TABLE 10.2

DISABILITY CATEGORY	SEX*	RACE/ETHNICITY					Optional** (6) TOTAL	EDUCATIONAL PLACEMENT: TIME OUTSIDE REGULAR CLASSROOM			(10) LEP****
		(1) American Indian or Alaskan Native	(2) Asian or Pacific Islander	(3) Hispanic	(4) Black, Not of Hispanic Origin	(5) White, Not of Hispanic Origin		(7) Less than 21%***	(8) Between 21% and 60%***	(9) More than 60%***	
A. Emotional Disturbance	M										
	F										
B. Specific Learning Disability	M										
	F										

Total of Table 10.2. See page 6 of instruction sheet

PLEASE FILL IN ALL INFORMATION INCLUDING LEP ITEMS. IF THE ANSWER IS ZERO, PLEASE ENTER ZERO. (For example, if you entered "school total" LEP data other than zero in Table 7 (Table 7, Row C, Column 6), but your school has no LEP students in the more specific "sub-total" categories on this page, please enter zero. If you have LEP students in these subcategories, please enter the number of these students.)

TABLE 10.3 IS TO BE COMPLETED ONLY BY SCHOOLS IN STATES WHICH HAVE ADOPTED THE CATEGORY OF DEVELOPMENTAL DELAY FOR STUDENTS WITH DISABILITIES IN KINDERGARTEN THROUGH AGE 9.

Table 10.3 Developmental Delay. Schools are *only* permitted to submit data using the following table for children with disabilities in kindergarten through age 9: if 1) their state has adopted this category in accordance with the provisions of Section 602(3)(b) of the *Individuals with Disabilities Education Act* (IDEA); 2) the school district completing the ED 101 form has adopted this category; and 3) the state is actually using this category to report data from this school district for IDEA child count purposes. *Do not provide data using this table unless all three of these IDEA requirements are met.*

TABLE 10.3

DISABILITY CATEGORY	SEX*	RACE/ETHNICITY					Optional** (6) TOTAL	EDUCATIONAL PLACEMENT: TIME OUTSIDE REGULAR CLASSROOM			(10) LEP****
		(1) American Indian or Alaskan Native	(2) Asian or Pacific Islander	(3) Hispanic	(4) Black, Not of Hispanic Origin	(5) White, Not of Hispanic Origin		(7) Less than 21%	(8) Between 21% and 60%***	(9) More than 60%***	
Developmental Delay	M										
	F										

* SEX: M = MALE; F = FEMALE

** Dark lines denote that this part of the table is optional. (See page 6 of instruction sheet.)

*** See Definitions on page 3 of ED102 instruction sheet for definitions, including complete label for *Less than 21%, Between 21% and 60%, and More than 60%*

**** LEP = limited English proficient

ORIGINAL - Return to Office for Civil Rights (LEGAL COPY)

17300

District Name: _____

School Name: _____

Individual School Report: ED102 — Page 5 of 10

Item 11. Additional Categories of Children with Disabilities: Please report by educational placement (the percentage of the day a student receives special education services outside the regular class), the additional children receiving special education services under the *Individuals with Disabilities Education Act.* Do not count *pre-kindergarten/pre-school* children. Include all students attending this school regardless of whether they are resident or non-resident of the reporting school district.

TABLE 11

DISABILITY CATEGORY	(1) LESS THAN 21% OF TIME OUTSIDE REGULAR CLASSROOM*	(2) BETWEEN 21% AND 60% OF TIME OUTSIDE REGULAR CLASSROOM*	(3) MORE THAN 60% OF TIME OUTSIDE REGULAR CLASSROOM*	(4) (Optional) TOTAL [COLUMN (1) + COLUMN (2) + COLUMN (3)]
A. Hearing Impairments				
B. Speech or Language Impairments				
C. Visual Impairments				
D. Orthopedic Impairments				
E. Autism				
F. Traumatic Brain Injury				
G. Deaf-Blindness				
H. Multiple Disabilities				
I. Other Health Impairments				
J. TOTAL (Optional. See Instructions)				

*See DEFINITIONS on page 3 of ED102 instruction sheet for definition, including the complete label.

CERTIFICATION: *I certify that the information is true and correct to the best of my knowledge and belief. A willfully false statement is punishable by law (U.S. Code, Title 18, Section 1001).*

Printed Name of Principal or Authorized Representative

Date

Telephone

Title

Fax

ORIGINAL - Return to Office for Civil Rights (LEGAL COPY)

Please Note:

1. Schools offering elementary grades - **Please complete Item 12A on Page 6 and Item 13 on Page 8 of the ED-102.**

2. Schools offering high school grades - **Please complete Item 12B on Page 7, Items 14 and 15 on Page 9 and Item 16 on Page 10 of the ED-102.**

3. Schools offering middle school grades - **Please complete Item 12A on Page 6 of the ED-102.**

4. **ALL SCHOOLS MUST COMPLETE ITEM 17 ON PAGE 10 OF THE ED-102.**

   See instructions regarding maintaining data by sex for students with disabilities.

NOTE: **PLEASE RETURN ALL SHEETS OF THE ORIGINAL FORMS TO THE OFFICE FOR CIVIL RIGHTS, EVEN IF YOU DID NOT USE ONE OF PAGES 6, 7, 8, OR 9 BECAUSE IT DID NOT PERTAIN TO YOUR SCHOOL.**

17300

District Name: _____

School Name: _____

Individual School Report: ED102 — Page 6 of 10

17300

TO BE COMPLETED BY ELEMENTARY AND MIDDLE SCHOOLS (GRADES K-8) ONLY

Item 12A. Testing (Grade to Grade Promotion). Please complete the following table if your school administered, in the 1999-2000 school year, a district- or state-required test that students are either required to pass or that is used as a significant factor in making promotion decisions for all students taking the test.
If your school conducted tests for grade-to-grade promotion for more than one grade, please photocopy the page (prior to completing) as many times as are necessary in order to report on each test, and report data using both the table on this page and as many photocopied tables as are appropriate. If students were **not** required to pass a district- or state-required test to be promoted from one grade to the next, please darken the bubble entitled "No such tests were administered". If students were required to pass such a test, please darken the appropriate bubble indicating whether this test was the "sole criterion" or a "significant" criterion and complete the table. If all students were required to take a district- or state-required test, and must pass the test to be promoted from one grade to the next, please **darken** the bubble entitled "Sole criterion". However, if all students were required to take the test, and the test is an important criterion in the decision on whether or not to promote the student from grade to grade, but other criteria, such as teacher recommendations or the student's grades were used in the promotion decision, please **darken** the bubble entitled "Significant criterion". Please provide the following data for the most recent testing of students in these grades during the 1999-2000 school year by race/ethnicity, limited English proficiency (in the column marked LEP), and whether the student is receiving services under the Individuals with Disabilities Education Act (in the column entitled Students with Disabilities/IDEA), or under Section 504 of the Rehabilitation Act of 1973 (in the column entitled Section 504 Only) and sex. Do not count students who were not tested because they had passed the test on a previous occasion. Include in Rows A or B those students who took the test and were provided with accommodations, modifications, or adaptations, such as a different setting, extended time, Braille, or use of dictionaries by LEP students. All students who did not take the test should be reported in Row C. Students who were tested using alternate assessments should be reported in Row D. An alternate assessment is an assessment provided to children with disabilities who cannot participate in a state- or district-wide assessment program, even with appropriate accommodations. If students are required to pass more than one test in order to be promoted from one grade to the next, include that student in the Row entitled Tested and Passed if that student passed all tests that he or she was required to pass; if that student failed one or more tests, report that student in the Row entitled Tested and Failed.

Please darken the appropriate bubble for information reported in this table:

○ No such tests were administered   ○ Sole criterion   ○ Significant criterion

Tests were required for promotion to:
○ Grade 1   ○ Grade 3   ○ Grade 5   ○ Grade 7
○ Grade 2   ○ Grade 4   ○ Grade 6   ○ Grade 8   ○ Grade 9

TABLE 12A

NUMBER OF STUDENTS	SEX*	RACE/ETHNICITY (1) American Indian or Alaskan Native	(2) Asian or Pacific Islander	(3) Hispanic	(4) Black, Not of Hispanic Origin	(5) White, Not of Hispanic Origin	(6) TOTAL	(7) Students with Disabilities/ IDEA	(8) Section 504 Only	(9) LEP**
A. Tested and passed	M									
	F									
B. Tested and failed	M									
	F									
C. Not tested	M									
	F									
D. Alternate Assessments	M									
	F									

* SEX:
M = MALE;
F = FEMALE

**LEP = limited English proficient

IMPORTANT! RETURN THIS PAGE EVEN IF IT WAS NOT FILLED OUT

ORIGINAL - Return to Office for Civil Rights (LEGAL COPY)

District Name: _____

School Name: _____

17300

TO BE COMPLETED BY HIGH SCHOOLS (GRADES 9-12) ONLY

Item 12B. Testing (High School Graduation). Please complete the following table *if your school administered, in the 1999-2000 school year, a district- or state-required test that students are either required to pass or that is used as a significant factor in making graduation decisions for all students taking the test.* If students were not required to pass a district- or state-required test to graduate from high school, please **darken** the bubble entitled *"No such tests were administered"*. If students were required to pass such a test, please darken the appropriate bubble indicating whether the test was the "sole criterion" or a "significant" criterion and complete the table. If all students were required to take a district- or state-required test, and must pass the test to graduate from high school, please **darken** the bubble entitled *"Sole criterion"*. However, if all students were required to take the test, and the test is an important criterion in the decision on whether or not the student graduates from high school, but other criteria, such as teacher recommendations or the student's grades were used in the graduation decision, please **darken** the bubble entitled *"Significant criterion"*. Please provide the following data for the testing of students in these grades during the 1999-2000 school year by race/ethnicity, limited English proficiency (in the column marked *LEP*), and whether the student is receiving services under the *Individuals with Disabilities Education Act* (in the column entitled *Students with Disabilities/IDEA*), or under *Section 504 of the Rehabilitation Act of 1973* (in the column entitled *Section 504 Only*) and sex. *Do not count students who were not tested because they had passed the test on a previous occasion.* Include in Rows A or B those students who took the test and were provided with accommodations, modifications, or adaptations, such as a different setting, extended time, Braille, or use of dictionaries by LEP students. All students who did not take the test and have not passed the test in prior years should be reported in Row C. Students who were tested using alternate assessments should be reported in Row D. An alternate assessment is an assessment provided to children with disabilities who cannot participate in a state- or district-wide assessment program, even with appropriate accommodations. If students are required to pass more than one test in order to graduate, include that student in the Row entitled *Tested and Passed* if that student passed all tests that he or she was required to pass; if that student failed one or more tests, report that student in the Row entitled *Tested and Failed.*

Please darken the appropriate bubble for information reported in this table:

○ **No such tests were administered**   ○ **Sole criterion**   ○ **Significant criterion**

TABLE 12B

NUMBER OF STUDENTS	S E X	RACE/ETHNICITY					(6) TOTAL	(7) Students with Disabilities/ IDEA	(8) Section 504 Only	(9) LEP**
		(1) American Indian or Alaskan Native	(2) Asian or Pacific Islander	(3) Hispanic	(4) Black, Not of Hispanic Origin	(5) White, Not of Hispanic Origin				
A. Tested and passed	M									
	F									
B. Tested and failed	M									
	F									
C. Not tested	M									
	F									
D. Alternate Assessments	M									
	F									

* SEX:
M = MALE;
F = FEMALE

**LEP = limited English proficient

IMPORTANT! RETURN THIS PAGE EVEN IF IT WAS NOT FILLED OUT

Appendix D

Using E&S Survey Data in Combination with Other Federal Datasets

Researchers with a license from the National Center for Education Statistics (NCES) to use the restricted versions of the Early Childhood Longitudinal Study (ECLS-K) and the National Educational Longitudinal Study (NELS:88) files can create new files using school-level information from the E&S survey for the appropriate years. Information from the Common Core of Data (CCD) is not restricted, so the school IDs are available to everyone. In this appendix, we discuss the potential benefits of merging E&S survey data with those from the CCD, ECLS-K, and NELS:88. Table D-1 lists the measures available from the E&S survey and indicates where comparable measures are available in the other datasets. At the very least, the table may provide a useful way to check the reliability of individual school-level measures. However, if there are discrepancies, we have no basis for determining which datafiles contain the more reliable data.

In addition to data about schools, both ECLS-K and NELS:88 collected data from individual teachers. With ECLS-K, sampled children's kindergarten and first grade teachers were surveyed. With NELS:88, information was collected from two of each sampled student's teachers in tenth grade, and from their math or science teachers in twelfth grade. Researchers might consider differences in how children's days are structured in elementary and high school when considering how to construct their analyses. In early elementary school, children's school experiences are limited largely to one teacher in one classroom, so the classroom as the unit of analysis makes good sense conceptually and statistically.[1] In

---

[1]There is no way to connect the classroom-level measures available on the E&S survey to the classrooms (and children) on ECLS-K. ECLS-K classroom-level measures are available only for the grades that have been sampled thus

high schools, however, students' experiences are spread over six or seven (or more) teachers each year, meaning that the appropriate unit of analysis is *not* the classroom but rather the school. In addition to cognitive assessments, NELS:88 also collected data from sampled students about their activities and attitudes with paper-and-pencil surveys. When appropriate, we note how students' responses might inform school-level investigations. All child-level reports of behaviors on ECLS-K are made by teachers or parents.

E&S survey data are available at both the district and the school levels, and Table D-1 shows the data that are available at the lowest level of aggregation, which is generally the school, but for some measures it is individual classrooms. Virtually all E&S data are broken down into multiple student categories (see the table footnotes). Throughout the table, the symbol * between two student characteristics indicates that the measure is broken down by more than one student characteristic (e.g., gender and race and

ethnicity). For example, "Total Student Enrollment Gender*Grade" indicates that the E&S survey provides information about the number of boys and girls in each grade. Similarly, "Grade*Race" indicates how many black eighth graders or white third graders are enrolled in a particular school.

The ECLS-K measures listed in Table D-1 are from the first grade file, measured in 1999–2000. NELS:88 measures are from the second follow-up, which occurred during the 1991–1992 school year, when most students were in twelfth grade. It is important that researchers planning to create combined E&S/NELS:88 data files use E&S survey data from the appropriate year: for example, the 1992 E&S survey data for those using data from the second NELS:88 follow-up. Much of the aggregate information available on the E&S survey that is unavailable at the school level on ECLS-K or NELS:88 is available about individual sampled students, although researchers should be judicious in making school-level aggregates from student-level data. Although within-school student samples were randomly drawn, the numbers for which some aggregates are drawn is small, and they are all in a single grade. Furthermore, many similar (and even identical) school-level measures are also available for the ECLS-K base year (when sampled children were in kinder-

---

far (kindergarten and first grade). When we indicate that data are available at the classroom level, this is in reference to sampled classrooms, and not each school's population of classrooms even for kindergarten and first grade.

garten) as well as the NELS:88 base year (eighth grade) and the first follow-up (tenth grade).

SCHOOL ENROLLMENT

The first category in the table includes measures in which the four datasets have the most in common. Each dataset includes measures indicating total school enrollment and school enrollment by race. With the CCD data, by "calculable" we mean simply adding two or more variables will produce a measure identical to that on the E&S survey. Information about enrollment by gender and race in each grade is also available on ECLS-K for sampled classrooms. Because the CCD contains no more information in common with the E&S survey, we do not mention it further in this section.

STUDENT DISCIPLINE, GIFTED AND TALENTED, AND ENGLISH AS A SECOND LANGUAGE

E&S contains information about the number of student disciplinary actions broken down by various student characteristics. Surprisingly, neither ECLS-K nor NELS:88 provides such information, although this information is available on NELS:88 on sampled students. ECLS-K includes the number of students in gifted and talented programs at both the school and sampled classroom levels. ECLS-K and NELS:88 also provide information about the number of students in classes for English-language learners in each school, while ECLS-K has information about the number of English-language learners in kindergarten and first grade. NELS:88 includes a measure indicating the number of children in each school receiving bilingual and English-language learner services. On ECLS-K, information about the number of children actually receiving such services is available only in sampled classrooms.

STUDENTS WITH DISABILITIES

The E&S survey distinguishes between two types of special education students: 504 and IDEA. The following definitions are included with E&S documentation:

> Section 504: An elementary or secondary student with a disability who is being provided with related aids and services under *Section 504* of the Rehabilitation Act of 1973, as amended, and is not being provided with services under the Individuals with Disabilities Education Act (IDEA).

IDEA: Under the Individuals with Disabilities Education Act (IDEA), children with mental retardation, hearing impairments including deafness, speech or language impairments, visual impairments including blindness, emotional disturbance, orthopedic impairments, autism, traumatic brain injury, other health impairments, or specific learning disabilities, deaf-blindness, multiple disabilities, or developmental delay; and who, by reason thereof, need special education and related services.

Because of the important federal role in special education, the E&S survey contains a great deal of information about special education access. NELS:88 provides simple information on the number of special education students enrolled in each school. ECLS-K includes only dichotomous measures indicating whether the school serves children with individual education plans (IEPs), children eligible under Section 504, and children with IEPs who are served under IDEA. However, in sampled classrooms, ECLS-K includes the number of children with IEPs and the number served under both IDEA and Section 504. The E&S survey also breaks down special education students into their specific disability; ECLS-K includes the same information, but only for sampled classrooms and does not distinguish degrees of mental retardation, as does the E&S survey.

Another area in which the E&S survey appears to be the sole source of school-level information is special education mainstreaming. E&S contains information about the number of students in each disability category that spend less than 21 percent, between 21 and 60 percent, and more than 60 percent of their time in regular classrooms. ECLS-K contains a very broad measure indicating whether special education students spend most of their day in or out of the regular classroom, as well as more detailed information on sampled students receiving special education services.

HIGH-STAKES TESTING

The E&S survey includes information about the number of students who passed or failed district- or state-mandated tests, as well as the number who were given alternative assessments or were simply not tested (again, broken down by student characteristics). ECLS-K includes only a measure indicating the proportion of students who performed at or above national norms on standardized tests of math and reading. NELS:88 indicates whether or not twelfth graders must pass a test to receive their high school diploma, whether students are required to pass minimum competency tests, and the percentage of students who fail these tests on the first attempt. Of course, testing was less of an issue a decade ago than it is today.

ABILITY GROUPING

Another set of E&S survey data details the number of students in each grade and class who are grouped by ability. Although quite interesting, there is no information regarding which children are in which groups (low, medium, or high), meaning that it is impossible to investigate whether within-school segregation by race, ethnicity, or gender is occurring. Information on these measures is included only for schools that have a 20–80 percent minority enrollment, suggesting that this area was a potential point of the survey item. However, the E&S survey measures do not provide a means to investigate such questions. One could certainly investigate whether schools that enroll different types of students tend to group their students by ability, but segregation by ability grouping is obviously a within-school phenomenon. Once ability groups are created, some segregation is likely to follow because test scores are stratified by race and social class throughout the K–12 system. Sampled ECLS-K classrooms contain data on whether a teacher sorts students by ability for math or reading,

but, again, there is no information about which children are in which groups.

TEACHERS AND HIGH SCHOOL DATA

The E&S survey includes the number of full-time equivalent (FTEs) teachers and the number of teachers who are fully certified in their subject area. Both ECLS-K and NELS:88 include the number of FTEs. In terms of teacher certification, ECLS-K and NELS:88 include such data on sampled teachers, which makes sense since teachers can be linked to students (and their achievement scores).

The E&S survey also collects information about student participation in advanced placement (AP) courses. The data include the number of students taking AP math and AP science courses. NELS:88 includes measures indicating the number of twelfth graders taking AP courses and the percent of the overall student body taking AP courses, and NELS:88 transcripts actually record student enrollment in AP courses. However, the NELS:88 data are likely to be outdated, as the AP program has mushroomed in the last decade. On the NELS:88 first follow-up (tenth grade), a measure is included indicating the number of students with limited English proficiency enrolled in AP classes.

Because of interest in Title IX compliance, the E&S survey includes information on the number of male and female sports each high school offers and the number of males and females participating. NELS:88 has no comparable measures other than data collected among sampled students regarding their participation in interscholastic activities.

SUMMARY

In some areas, such as the number and characteristics of students and teachers and the raw number of English-language learners and special education students in each school, data on the E&S survey are available on other federal datasets. In several other areas, however, such as student discipline and special education mainstreaming, E&S appears to be an important source of national data and is certainly the only dataset containing such information on the entire population of U.S. public schools, and over time, for the same schools. For researchers interested in national trends in the programs surveyed by the E&S survey, the data are an invaluable resource, especially because individual schools have been surveyed for almost 35 years. The potential to investigate change in these schools' compliance is enormous. However, for researchers interested in school effects studies involving student-level social and academic outcomes, the E&S data are less valuable.

TABLE D-1 Availability of E&S Survey Measures in CCD, ECLS-K, and NELS:88 Datasets

E&S Measures: Number of Students	CCD Measures
School Enrollment	
Enrollment	MEMBER00
Enrollment by grade[a]	G0100, G0200, etc.
Enrollment by race	ASIAN00, HISP00, BLACK00, WHITE00, AM00,
Enrollment grade*race[a]	calculable
Enrollment by gender	calculable
Enrollment gender*grade[a]	calculable
Student Discipline[c]	
Receiving corporal punishment	—
Receiving out-of-school suspensions	—
Expelled	
—	—
Gifted/Talented and Language Services	
In gifted and talented programs[d]	—
In gifted and talented programs by grade	—
Needing LEP programs[e]	—
Enrolled in LEP programs[f]	—
Students with Disabilities	
Under Section 504	—
Under IDEA	—
Total mental retardation[g]	—
Mild mental retardation[g]	—
Moderate mental retardation[g]	—
Severe mental retardation[g]	—
Emotional disturbance	—
Specific learning disabilities	—
Developmental delays	—
Hearing impairments	—

ECLS-K Measures	NELS:88 Measures
S4ENRLS	F2SCENRL
S4ENRLK, S4ENRLF	G12ENROLL (12th gr. enrol.)
S4ASNPCT, S4HSPPCT, S4BLKPCT, S4WHTPCT, S4INDPCT, S4OTHPCT	F2C22A-F2C22E
A4ASIAN, A4HISP, etc.[b]	—
—	—
A4BOYS, A4GIRLS[b]	—
—	—
—	—
—	
S4GFTNBR	—
A4PRTGF[2]	—
S4LEPSCH; S4LEPFRS (% 1st grade)	F2C24
A4NUMLE[b]	
Calculable using A4ESLRE, A4ESLOU, and	F2C25F (Bilingual)
A4NOESL[b]	F2C25G (ESL)
S4ON504 (enroll these students, yes or no)	
A4SC504[b]	—
S4ONIEP, S4IEP504 (enroll these students, yes or	F2C25H
no) A4IEP[b]	
A4RETAR[b]: no distinction between degree of	
retardation.	
—	—
—	—
—	—
A4EMPRB[b]	—
A4LRNDI[b]	—
A4DELAY[b]	—
A4HEAR[b]	—

E&S Measures: Number of Students	CCD Measures
Speech or language impairments	— A4IMP[b]
Visual impairments	— A4VIS[b]
Orthopedic impairments	— A4ORTHO[b]
Autism	— A4AUTISM[b]
Traumatic brain injury	— A4TRAUM[b]
Deaf or blind	— A4DEAF[b]
Multiple disabilities	— A4MULTI[b]
Other health impairments	— A4OTHER[b]
Mainstreaming[h]	
Total mental retardation	—
Mild mental retardation	—
Moderate mental retardation	—
Severe mental retardation	—
Emotional disturbance	—
Specific learning disabilities	—
Developmental delays	—
Hearing impairments	—
Speech or language impairments	—
Visual impairments	—
Orthopedic impairments	—
Autism	—
Traumatic brain injury	—
Deaf or blind	—
Multiple disabilities	—
Other health impairments	—
High-Stakes Testing	
Passing, failing, given alternative assessments, or not tested for a district- or state-administered test for promotion or graduation[j]	—
Ability Grouped Classrooms (through 8th grade) [a,k]	
	—
Teachers: Number	
Fulltime	—
Certified fulltime	—

ECLS-K Measures	NELS:88 Measures
—	
—	
—	
—	
—	
—	
—	
—	
S4DISSRV[i]	—
—	—
—	—
—	—
—	—
—	—
—	—
—	—
—	—
—	—
—	—
—	—
—	—
—	—
—	—
—	—
% perform at or above grade level national norms in math (S4PCTMTH)and reading (S4PCTRD):	F2C42: seniors must pass test for high school diploma F2C43A-F2C43F: minimum competency tests req. 7-12 F2C46: % initially fail tests
Ability groups for reading (A4DIVRD) or math (A4DIVMTH)[b]	—
S4FTETOT TB4TYPCE[b] (classroom level)	F2C29A F2T4_7A: certification of sampled math teachers F2T4_7B: certification of sampled science teachers

TABLE D-1 Continued

E&S Measures: Number of Students	CCD Measures
High School Specific	
Taking AP math or science courses[f]	—
Participating in interscholastic athletic activities by gender, and number of sports and number of teams for males and females	—

NOTES: Datasets: CCD, Common Core of Data; ECLS-K, Early Childhood Longitudinal Study, Kindergarten and First grade; NELS:88, National Educational Longitudinal Study, 1988. For each dataset (column), the entries are the variable names in that dataset. An asterisk (*) indicates the number is available by two or more variables. A dash (—) indicates the information is not available in the dataset. LEP, students with limited English proficiency.

[a]This information available only for schools enrolling 20–80 percent minority students.
[b]Data are available only for sampled classrooms.
[c]Data also sorted by race, gender, IDEA, 504, LEP, race*gender, LEP*gender, IDEA*gender, and 504*gender.
[d]Data also sorted by race, gender, LEP, IDEA, race*gender, LEP*gender, and IDEA*gender.
[e]Data also sorted by race, gender, IDEA, LEP, race*gender, IDEA*gender, LEP*gender, grade, classroom, and grade*classroom.
[f]Data also sorted by race, gender, IDEA, LEP, race*gender, IDEA*gender, and LEP*gender.

ECLS-K Measures	NELS:88 Measures
—	F2C49: # 12th graders in AP courses
	FC25I: % students who take AP courses
—	Sampled students only

[g]Data also sorted by race, gender, LEP, race*gender, and LEP*gender.

[h]Mainstreaming is indicated by the number of students with this disability who are in a regular classroom <21 percent of the time, between 21–60 percent of the time, and more than 60 percent of the time. These data are also sorted by gender.

[i]Measure simply indicates whether children with disabilities spend most of their day in or out of the regular classroom.

[j]Data also sorted by grade (K–8 only), race, gender, 504, IDEA, LEP, grade* gender, 504*gender, IDEA*gender, LEP*gender, race*grade, race*grade*gender, grade*gender*504, grade*gender*IDEA, grade*gender*LEP, grade*504, and grade*504.

[k]Data also sorted by grade, classroom, race, LEP, race*classroom, LEP*classroom, race*grade, LEP*grade, grade*classroom, race*grade*classroom, and LEP*grade*classroom.

[l]Data also sorted by race, gender, IDEA, LEP, race*gender, IDEA*gender, LEP*gender.